Engineering Optimization: Methods and Applications

Optimization carries great significance in both human affairs and the laws of nature. It refers to a positive and intrinsically human concept of minimization or maximization to achieve the best or most favorable outcome from a given situation. Besides, as the resources are becoming scarce there is a need to develop methods and techniques which will make the systems extract maximum from minimum use of these resources, i.e. maximum utilization of available resources with minimum investment or cost of any kind. The resources could be any, such as land, materials, machines, personnel, skills, time, etc. The disciplines such as mechanical, civil, electrical, chemical, computer engineering as well as the interdisciplinary streams such as automobile, structural, biomedical, industrial, environmental engineering, etc. involve in applying scientific approaches and techniques in designing and developing efficient systems to get the optimum and desired output. The multifaceted processes involved are designing, manufacturing, operations, inspection and testing, forecasting, scheduling, costing, networking, reliability enhancement, etc. There are several deterministic and approximation-based optimization methods that have been developed by the researchers, such as branch-and-bound techniques, simplex methods, approximation and Artificial Intelligence-based methods such as evolutionary methods, Swarm-based methods, physics-based methods, socio-inspired methods, etc. The associated examples are Genetic Algorithms, Differential Evolution, Ant Colony Optimization, Particle Swarm Optimization, Artificial Bee Colony, Grey Wolf Optimizer, Political Optimizer, Cohort Intelligence, League Championship Algorithm, etc. These techniques have certain advantages and limitations and their performance significantly varies when dealing with a certain class of problems including continuous, discrete, and combinatorial domains, hard and soft constrained problems, problems with static and dynamic in nature, optimal control, and different types of linear and nonlinear problems, etc. There are several problem-specific heuristic methods are also existing in the literature.

This series aims to provide a platform for a broad discussion on the development of novel optimization methods, modifications over the existing methods including hybridization of the existing methods as well as applying existing optimization methods for solving a variety of problems from engineering streams. This series publishes authored and edited books, monographs, and textbooks. The series will serve as an authoritative source for a broad audience of individuals involved in research and product development and will be of value to researchers and advanced undergraduate and graduate students in engineering optimization methods and associated applications.

Ritu Arora · Shalini Arora
Editors

Optimization Techniques for Sustainable Environment Under Uncertainty

Springer

Editors
Ritu Arora
Department of Mathematics
Keshav Mahavidyalaya
University of Delhi
New Delhi, Delhi, India

Shalini Arora
Department of Applied Sciences
and Humanities
Indira Gandhi Delhi Technical University
for Women
New Delhi, Delhi, India

ISSN 2731-4049 ISSN 2731-4057 (electronic)
Engineering Optimization: Methods and Applications
ISBN 978-981-96-0388-6 ISBN 978-981-96-0389-3 (eBook)
https://doi.org/10.1007/978-981-96-0389-3

This Springer imprint is published by the registered company Springer Nature Singapore Pte Ltd.
The registered company address is: 152 Beach Road, #21-01/04 Gateway East, Singapore 189721, Singapore

Contents

Editors and Contributors

About the Editors

Dr. Ritu Arora is a Professor in the Department of Mathematics at Keshav Mahavidyalaya, University of Delhi. She has over 20 years of teaching experience and earned her Masters and Ph.D. in Mathematics from the University of Delhi. Her research focuses on Mathematical Programming/Optimization and its application to various allocation problems. Dr. Arora is the author of the book *Linear Programming and Game Theory* Narosa Publications and has published over 25 research papers in reputable National and International Journals.

Dr. Shalini Arora is a Professor in the Department of Mathematics and Head of the Applied Sciences and Humanities Department at Indira Gandhi Delhi Technical University for Women (IGDTUW). With over 20 years of teaching experience, she completed her Master's and Ph.D. in Mathematics from IIT Delhi. Her research interests include Mathematical Programming, Allocation Problems such as Transportation and Assignment Problems, Combinatorial optimization, and more. Dr. Arora has published over 30 research papers in both International and national journals. She is a recipient of the 'Young Scientist Award' by the SERC division of DST.

Contributors

Parastoo Afrasyabi Faculty of Geodesy and Geomatics Engineering K.N, Toosi University of Technology, Tehran, Iran

Divya Agarwal Amity Institute of Applied Sciences, Amity University, Noida, UP, India

Radhika Agarwal Amity Institute of Applied Sciences, Amity University, Noida, UP, India

P. Anukokila Department of Mathematics, PSG College of Arts & Science, Coimbatore, TN, India

Ritu Arora Department of Mathematics, Keshav Mahavidyalaya, University of Delhi, Delhi, India

Shalini Arora Department of Applied Sciences and Humanities, Indira Gandhi Delhi Technical University for Women, Delhi, India

Arpita Bhowmik Department of Applied Mathematics, Vidyasagar University, Midnapore, India

Aryan Bhusari Institute of Artificial Intelligence, MIT World Peace University, Pune, Maharashtra, India

Salar Farahmand-Tabar Faculty of Engineering, Department of Civil Engineering, University of Zanjan, Zanjan, Iran

Savita Garg Department of Mathematics, Mukand Lal National College, Haryana, India

Akshada Ghoderao Department of Electronics and Telecommunication Engineering, MIT Academy of Engineering, Pune, Maharashtra, India

Shreyansh Khokale Department of Electronics and Telecommunication Engineering, MIT Academy of Engineering, Pune, Maharashtra, India

Priyanshi Kotian Department of Electronics and Telecommunication Engineering, MIT Academy of Engineering, Pune, Maharashtra, India

Anand J. Kulkarni Institute of Artificial Intelligence, MIT World Peace University, Pune, Maharashtra, India

Shruti Nare Department of Electronics and Telecommunication Engineering, MIT Academy of Engineering, Pune, Maharashtra, India

Madhumangal Pal Department of Applied Mathematics, Vidyasagar University, Midnapore, India

Savita Pawar Department of Electronics and Telecommunication Engineering, MIT Academy of Engineering, Pune, Maharashtra, India

B. Radhakrishnan Department of Mathematics, PSG College of Technology, Coimbatore, Tamil Nadu, India

Prachi Rajarapollu Department of Electronics and Telecommunication Engineering, MIT Academy of Engineering, Pune, Maharashtra, India

Neetu Rani Department of Mathematics, Shivaji College (University of Delhi), Delhi, India

Laxminarayan Sahoo Department of Computer and Information Science, Raiganj University, Raiganj, India

Sovan Samanta Department of Mathematics, Tamluk, West Bengal, India; Department of Technical Sciences, Western Caspian University, Baku, Azerbaijan

Punita Saxena Department of Mathematics, Shaheed Rajguru College of Applied Sciences for Women, Delhi, India

Bharti Seth School of Engineering and Technology, Vivekananda Institute of Professional Studies, Delhi, India;
Department of Applied Science and Humanities, Indira Gandhi Delhi Technical University for Women, Delhi, India

Sunny Seth Jagan Institute of Management Studies, New Delhi, India

Vivek Kumar Sharma JaganNath University, Jaipur, Rajasthan, India

Aakanksha Singh Department of Applied Science and Humanities, Indira Gandhi Delhi Technical University for Women, Delhi, India;
Department of Mathematics, Aryabhatta College, University of Delhi, Delhi, India

Bhupender Kumar Som GNIOT Institute of Management Studies, Uttar Pradesh, Greater Noida, India

Chapter 1
An Empirical Study of Indian Passenger Airlines for Sustainable Performance: A Window DEA Approach

Bharti Seth, Punita Saxena, and Shalini Arora

Abstract The present study aims to examine and contrast the operational efficacy of scheduled passenger airlines operating in India from 2014 to 2021. Conventional Data Envelopment Analysis (DEA) models determine efficiency scores (technical, pure technical, and scale). The performance of airlines is tracked throughout time with the aid of window analysis. A mix of both will help evaluate the airlines between 2014 and 2021. Aiming to help decision-makers improve their organizations' performance for long-term, sustainable growth, improvements and projections are investigated.

Keywords Data envelopment analysis · Indian airlines · Window analysis · Performance assessment

1.1 Introduction

A well-knit and coordinated transport system is essential for the country's sustainable growth. The transport system forms the economy's veins as it is the foundation of the country's economic infrastructure. The Indian transportation sector contributes to a country's overall financial performance. The network spread and output of the transportation system have expanded exponentially over the years.

The excellent aviation business in India is currently in operation, with expansive networks and various functions. As per the press release on February 2, 2023, Press Release, India has witnessed a double-digit Compound Annual Growth Rate (CAGR)

B. Seth (✉)
School of Engineering and Technology, Vivekananda Institute of Professional Studies, Delhi, India
e-mail: bharti.seth@vips.edu

B. Seth · S. Arora
Department of Applied Sciences and Humanities, Indira Gandhi Delhi Technical University for Women, Delhi, India
e-mail: shaliniarora@igdtuw.ac.in

P. Saxena
Department of Mathematics, Shaheed Rajguru College of Applied Sciences for Women, Delhi, India

R. Arora and S. Arora (eds.), *Optimization Techniques for Sustainable Environment Under Uncertainty*, Engineering Optimization: Methods and Applications,
https://doi.org/10.1007/978-981-96-0389-3_1

of 14.5% in domestic passenger traffic during 2014–2015 to 2019–2020. Domestic aviation traffic decreased in 2020–2021 and 2021–2022 due to the COVID-19 pandemic. However, it has resumed its upward trend for the fiscal year 2022–2023 and is expected to reach approximately 97% of its pre-Covid level. The government is actively assisting the aviation industry by fostering a stable policy environment and encouraging growth driven by competition.

One of the key concerns for decision-makers is efficiency measurement. They cannot discover the performance of the units under their leadership without a research methodology due to the complexity of the information, the volume of data, and the influence of numerous other factors. The empirical investigation in this paper will help to understand the effect of variables under study on the operational performance of the airlines operating in India. The efficiency of the units being evaluated is one of the core elements in performance evaluation. Performance evaluation can be either achieved by non-parametric or parametric approaches.

Non-parametric data envelopment analysis (DEA) is based on the framework of Farrell (1957), which was further expanded by Charnes et al. (1978) and Banker et al. (1984). DEA has an advantage over other mathematical programming and statistical techniques in the intricacy of the interactions between various inputs and outputs, where the units are typically not comparable. DEA is a non-parametric mathematical programming technique used to compare the observed with the decision-making unit's desired level of working performance ratio. The performance ratio is bounded by zero and unity. In constructing an efficient frontier, DMU_S are categorized into two groups. One that lies on the efficient frontier has an efficiency score equal to one, and those whose efficiency scores are less than one lie off to the frontier. Window analysis DEA technique is used for the dynamic performance appraisal of the Indian aviation sector during different periods between the time windows for 2014–2021.

1.2 Conventional DEA Models and Approaches

Standard DEA models, usually referred to as radial efficiency measures, are termed conventional models. They can either be input- or output-focused. The Charnes, Cooper, and Rhodes (CCR) model is the most popular DEA model studied and described in the literature (Charnes et al. 1978). An extension of the CCR model is known as the Banker, Charnes, and Cooper (BCC) model (Banker et al. 1984). The CCR model was based on the assumption that the relationship between inputs and outputs would have constant returns to scale (CRS). The BCC model was obtained by adding a variable returns to scale (VRS) ratio between the inputs and outputs. For the CCR model, input-oriented and output-oriented formulations produce the same results, but this is not the case for the BCC model. The input-oriented model attempts to minimize input quantities while maintaining at least the current output levels. The output-oriented model aims to maximize the output levels while using no more input than necessary.

1.2.1 CCR Model

Consider a set of 'n' DMU_S with input and output matrices X and Y respectively.

$$X = \begin{bmatrix} x_{11} & x_{12} & \cdots & x_{1n} \\ x_{21} & x_{22} & \cdots & x_{2n} \\ . & . & \cdots & . \\ . & . & \cdots & . \\ x_{r1} & x_{r2} & \cdots & x_{rn} \end{bmatrix}$$

$$Y = \begin{bmatrix} y_{11} & y_{12} & \cdots & y_{1n} \\ y_{21} & y_{22} & \cdots & y_{2n} \\ . & . & \cdots & . \\ . & . & \cdots & . \\ y_{s1} & y_{s2} & \cdots & y_{sn} \end{bmatrix}$$

For each jth DMU, $j = 1, 2, ..., n$, x_{ij}, $i = 1, 2, ..., r$, and y_{kj}, $k = 1, 2, ..., s$ are the r input and s output column vectors respectively. The efficiency of the target unit j_O is calculated using the following fractional programming model:

$$\left\{\begin{array}{l} Max\ \frac{\sum_{k=1}^{s} u_k y_{kj_O}}{\sum_{i=1}^{r} v_i x_{ij_O}} \\ s.t:\ \frac{\sum_{k=1}^{s} u_k y_{kj}}{\sum_{i=1}^{r} v_i x_{ij}} \le 1,\ j = 1, 2, \ldots, n \\ u_k, v_i > \epsilon, \forall k = 1, 2, \ldots, s, \forall i = 1, 2, \ldots, r \end{array}\right\} \tag{1.1}$$

where v_i is the weight of the ith input, u_k is the weight of the kth output.

Each DMU is given the flexibility to choose the weights for its factors to maximize its efficiency while ensuring that the efficiency of no other unit exceeds 1. Also, to prevent units from treating any element as an utterly free commodity and being mistakenly ranked as efficient, ϵ, an infinitesimal number has been introduced (Miliotis 1992). The non-Archimedean positive constant ϵ ensures no input or output is neglected while evaluating the efficiencies. The model (1.1) is solved by converting it to the linear model (1.2). The weights are generally referred to as multipliers. Thus, the model (1.2) is usually called the multiplier model, and its dual model (1.3) is called the envelopment model:

$$\left\{\begin{array}{l} Max \sum_{k=1}^{s} u_k y_{kj_O} \\ \text{subject to} \\ \sum_{i=1}^{r} v_i x_{ij_O} = 1 \\ \sum_{k=1}^{s} u_k y_{kj} \le \sum_{i=1}^{r} v_i x_{ij},\ j = 1, 2, \ldots, n \\ u_k, v_i > \epsilon, \forall k = 1, 2, \ldots, s, \forall i = 1, 2, \ldots, r \end{array}\right\} \tag{1.2}$$

Let $\lambda_1, \lambda_2,\ldots,\lambda_n$ be the n dual variables which represent the n DMU$_S$ and help to identify the benchmarks for the DMU$_S$ which fail to form the efficient frontier. Let ϕ_0 represent the technical efficiency (TE) of the corresponding DMU being assessed, and if the optimal value of ϕ_0 equals one, then the unit is termed technically efficient:

$$\left\{\begin{array}{l} Min\ \phi_0 \\ \text{subject to} \\ \lambda_j x_{ij} \le \phi x_{ij_0},\ j = 1, 2, \ldots, n \\ \lambda_j y_{kj} \ge y_{kj_0},\ j = 1, 2, \ldots, n \\ \lambda_j \ge 0 \end{array}\right\} \tag{1.3}$$

The set of feasible solutions for CCR models is called the production possibility set P_C defined by $P_C = \{(x, y)|x > X\lambda, y < Y\lambda, \lambda \ge 0\}$, where λ is a semi-positive vector in R^n. The models (1.1), (1.2), and (1.3) are referred to as input-oriented and constant returns to scale (CRS) DEA models since it is desirable to reduce the inputs while maintaining the output levels. Further, the CRS property implies if $(x, y) \in P_C$, then $(tx, ty) \in P_C$. Output-oriented models can also be formulated similarly. These concentrate on potential output growth by maintaining present input levels (Cooper and Seiford 2011).

In certain situations, the inefficiency in a unit can be due to its scale of operations. According to Banker and Thrall (1992), if $(\lambda_1^*, \lambda_2^*, \ldots, \lambda_n^*)$ is the optimal solution for the CCR model in envelopment form, then this model can also determine whether the returns to scale is increasing or decreasing at (x_0, y_0) point on the efficient frontier as per the following conditions:

- Constant returns to scale (CRS) prevails if $\sum_{j=1}^{n} \lambda_j^* = 1$ for any alternate optimum.
- Increasing returns to scale (IRS) prevails if $\sum_{j=1}^{n} \lambda_j^* < 1$ for all alternate optima.
- Decreasing returns to scale (DRS) prevails if $\sum_{j=1}^{n} \lambda_j^* > 1$ for all alternate optima.

1.2.2 BCC Model

Banker-Charnes-Cooper (Banker et al. 1984) modified the CCR model to formulate the BCC model. If the constraint $\sum_{j=1}^{n} \lambda_j = 1$ is added to the constraints, then the models (1.1), (1.2), and (1.3) are referred to as BCC models that assume variable returns to scale (VRS) frontiers, i.e. constant, increasing and decreasing returns to scale. If a rise in inputs does not cause a proportionate rise in outputs, then DMU runs under variable returns to scale. The set of feasible solutions for BCC models is called the production possibility set P_B defined by $P_B = \{(x, y)|x > X\lambda, y < Y\lambda, \lambda \ge 0, e\lambda = 1\}$, where λ is a semi-positive vector in R^n, and e is a row vector with all elements equal to 1.

The envelopment form model (1.4) and multiplier form model (1.5) of the BCC model are as follows:

$$\left\{\begin{array}{l} Min\ \phi_0 \\ \text{subject to} \\ \lambda_j x_{ij} \leq \phi x_{ij_0},\ j = 1, 2, \ldots, n \\ \lambda_j y_{kj} \geq y_{kj_0},\ j = 1, 2, \ldots, n \\ \sum_{j=1}^{n} \lambda_j = 1 \\ \lambda_j \geq 0 \end{array}\right\} \tag{1.4}$$

$$\left\{\begin{array}{l} Max\ \sum_{k=1}^{s} u_k y_{kj_0} - a_f \\ \text{subject to} \\ \sum_{i=1}^{r} v_i x_{ij_0} = 1 \\ \sum_{k=1}^{s} u_k y_{kj} \leq \sum_{i=1}^{r} v_i x_{ij} + a_f,\ j = 1, 2, \ldots, n \\ u_k, v_i > \epsilon, \forall k = 1, 2, \ldots, s, \forall i = 1, 2, \ldots, r \\ a_f \text{ is unrestricted.} \end{array}\right\} \tag{1.5}$$

Here, a_f is unrestricted dual variable and associated with the constraint $\sum_{j=1}^{n} \lambda_j = 1$ of the model (1.4). According to Banker and Thrall (1992), the following conditions at (x_0, y_0) point on the efficient frontier identify the returns to scale situations:

- Constant returns to scale (CRS) prevails if and only if $a_f^* = 0$ in any optimal solution.
- Increasing returns to scale (IRS) prevails if and only if $a_f^* < 0$ for all optimal solutions.
- Decreasing returns to scale (DRS) prevails if and only if $a_f^* > 0$ for all optimal solutions.

The CCR score is global technical efficiency because it is based on the constant returns-to-scale production possibility set, which holds that all observed DMU_S and their non-negative combinations can undergo radial expansion and reduction. On the other hand, the BCC model assumes that the convex hull of the DMU_S constitutes the production possibility set, and hence, the BCC score is called local pure technical efficiency. Due to the DMU_S scale size, if a DMU has BCC efficiency equal to one but a low CCR score, it is working locally efficiently but not globally efficiently. Thus, it is reasonable to characterize the scale efficiency of a DMU by the ratio of the two scores.

Thus, the technical efficiency (TE) obtained from the CCR model is divided into pure technical efficiency (PTE) and scale efficiency (SE). SE = $\frac{\text{TE}}{\text{PTE}}$ quantifies how the scale size impacts efficiency (Refaie et al. 2015). In both models, the units that form the frontier are the efficient units. These units alone or a combination of a few of them become the benchmarks for each of the inefficient units under study. The reference set or the peer group to the inefficient DMU is the set of collection of indices of efficient DMU_S. Existence of this reference set forces the unit under study to be inefficient.

1.2.3 Window Analysis

Comparing the performance of DMU_S over time is generally referred to as time series analysis. The variation in the efficiency of DMU_S over time can help the decision-makers analyze their performance more meaningfully. Window analysis is one methodology of using DEA in a time series mode. This model is a moving average pattern of analysis where each time period is termed a window. It helps study the performance trend of a unit over time. The performance of DMU is compared with its performance in other periods, in addition to comparing it with the performance of other DMU_S in the same period (Refaie et al. 2015). By doing so, the analysis includes more units, which is beneficial when working with fewer units under study. This approach implicitly presupposes no technical changes within any window because all units within a specific window are measured against one another. According to Asmild et al. (2004), this issue can be solved by using a small window width, and for the results of a DEA window analysis to be reliable, the window width should be selected so that it is acceptable to presume that each window's technical change is minimal.

Suppose that 'n' DMU_S are observed in τ periods. Let ${DMU_j}^t$ represent the unit observed in period t with input and output vectors as $X_j^t = [x_j^{1t}, x_j^{2t}, \ldots, x_j^{rt}]$ and $Y_j^t = [y_j^{1t}, y_j^{2t}, \ldots, y_j^{st}]$ respectively. Let the window initialize at the time T $(1 \leq T \leq \tau)$ with the window length $l\,(1 \leq l \leq (\tau - T))$, then the number of observations in each window will be $n * l$ and the input and output matrices are respectively represented as follows:

$$X_{Tl} = \begin{bmatrix} X_1^T & X_2^T & \ldots & X_n^T \\ X_1^{T+1} & X_2^{T+1} & \ldots & X_n^{T+1} \\ . & . & \ldots & . \\ . & . & \ldots & . \\ X_1^{T+l} & X_2^{T+l} & \ldots & X_n^{T+l} \end{bmatrix}$$

$$Y_{Tl} = \begin{bmatrix} Y_1^T & Y_2^T & \ldots & Y_n^T \\ Y_1^{T+1} & Y_2^{T+1} & \ldots & Y_n^{T+1} \\ . & . & \ldots & . \\ . & . & \ldots & . \\ Y_1^{T+l} & Y_2^{T+l} & \ldots & Y_n^{T+l} \end{bmatrix}$$

The DEA models for these input and output matrices will produce the window analysis results for ${DMU_j}^T$ (Refaie et al. 2015).

Table 1.1 Variables used to assess the performance of airlines for 2014–2021

Variables	Definition
$I_1 = \frac{\text{Operating Expense (In Millions)}}{\text{Fleet Size}}$	It includes all the operating expenses which an airline company is outlaying for the smooth functioning of its unit airline per year
$I_2 = \frac{\text{Expense on Personnel (In Millions)}}{\text{Number of Employees}}$	It represents the amount on each personnel an airline is spending, including their salaries, wages, bonuses, and other allowances per year
$O_1 = \frac{\text{RPK (In Millions)}}{\text{ASK (In Millions)}}$	It establishes the capacity utilization of any airline company. It is calculated by dividing the revenue passenger kilometers by the available seat kilometers
O_2 = Revenue	It comprises the money made from revenue-paying passengers and fees for additional baggage

1.3 Data and Variables

The data set consists of a group of 12 airlines operating in India owned by the government and private operators, namely Air India, Air India Express, Alliance Air, Air Asia, Go Air, Indigo, Jet Airways, Jet Lite, Spice Jet, True Jet, Vistara, and Star Air. Data related to efficiency indicators of scheduled airlines of both service types domestic and international for the period 2014–2021 have been collected from the website of the Office of the Directorate General of Civil Aviation (**DGCA**).[1] The economic structure, airline administrative procedures, and market performance substantially influence the operational success of an airline Saranga and Nagpal (2016). Thus, two input and two output variables for performance evaluation are defined in Table 1.1 based on these drivers. These variables were also employed by Seth et al. (2023), Saranga and Nagpal (2016), Kottas et al. (2018), and Joo and Fowler (2014) in their research. Table 1.2 presents the descriptive statistics for the data used to assess the performance of airlines for the period 2014–2021.

1.4 Results and Discussion

1.4.1 Efficiency Scores Using Conventional Models

The evaluation of efficiency scores using the traditional DEA models (CCR and BCC) forms an essential basis for analyzing the operational efficiencies of the airlines.

[1] DGCA AIR STATISTICS REPORTS.

Table 1.2 Descriptive statistics of the data for the period 2014–2021

		Maximum	Minimum	Average	Std. deviation
2014–2015	I_1	2246.08	303.4	1360.12	618.32
	I_2	2.04	0.85	1.23	0.31
	O_1	82.4	53.7	75.64	8.36
	O_2	165895.9	556.8	57450.50	63902.396
2015–2016	I_1	1858.63	292.24	1239.332	445.3183
	I_2	1.9131	0.69	1.31	0.41
	O_1	90.6	66.5	79.482	6.516
	O_2	181661.2	645.1	56013.17	67378.142
2016–2017	I_1	2039.135	24.7	1205.801	572.0002
	I_2	2.534	0.592	1.4113	0.54735
	O_1	91.6	68.8	79.817	6.5598
	O_2	182434	14	56388.205	68998.15335
2017–2018	I_1	2160.221	27.665	1092.008	677.57
	I_2	2.68	0.031	1.292	0.7267
	O_1	93.6	27.2	76.8214	16.4856
	O_2	213423.37	0.48	57086.574	76528.4518
2018–2019	I_1	2358.9133	20.9	1110.077	699.4942
	I_2	4.2985	0.212	1.602	1.13
	O_1	91.9	48.5	76.573	13.26
	O_2	269473.6	3	64975.591	86639.44
2019–2020	I_1	2548.891	177.071	1234.923	661.034
	I_2	3.54	0.6553	1.6106	0.824
	O_1	89.6	65.3	79.99	7.6360
	O_2	338285.33	478.89	87796.755	105151.025
2020–2021	I_1	1203.1	184.1654	672.2314	297.3734
	I_2	2.5703	0.38846	1.1186	0.63201
	O_1	76.8	51.1	65.295	7.327
	O_2	125416.66	665.27	30574.161	37755.68

Table 1.3 contains the efficiency scores using the conventional DEA models for each year under study. Conventional efficiency scores include the CCR efficiency or technical efficiency (TE). Since CCR model yields the same efficiency score with input- and output-oriented models, therefore only CCR scores are reflected in Table 1.3. BCC model is used to calculate purely technical efficiency (PTE) with respect to both input and output orientations.

CCR scores indicate that the performance of most airlines is declining over the period of study. Also, for seven airlines, BCC-I scores are lower than BCC-O scores. It indicates that airline companies cannot efficiently utilize the input resources while maintaining the same output level.

Table 1.3 Efficiency scores for the conventional models

		2014–2015	2015–2016	2016–2017	2017–2018	2018–2019	2019–2020	2020–2021
Air India	CCR	1	0.768	0.6142	0.53	0.44	0.37	0.4
	BCC-I	1	0.7737	0.628	0.53	0.444	0.37	0.41
	BCC-O	1	0.90882	0.912	0.905	0.904	0.91	0.91
Air India Express	CCR	0.8398	0.47	0.42	0.20	0.24	0.33	0.38
	BCC-I	1	0.62	0.42	0.41	0.47	0.481	0.411
	BCC-O	1	0.91	0.83	0.81	0.87	0.93	0.85
Alliance Air	CCR	1	1	0.53	0.21	0.12	0.39	0.37
	BCC-I	1	1	0.56	0.63	0.75	0.41	0.51
	BCC-O	1	1	0.88	0.97	0.92	0.87	0.69
Air Asia	CCR	0.82	0.54	0.65	0.17	0.3	0.55	0.31
	BCC-I	0.912	0.54	0.77	0.70	0.54	0.60	0.35
	BCC-O	0.97	0.89	0.93	0.92	0.92	0.94	0.83
Go Air	CCR	1	0.66	0.73	0.3	0.36	0.47	0.8
	BCC-I	1	0.68	0.80	0.73	0.76	0.73	0.86
	BCC-O	1	0.92	0.96	0.95	0.95	0.97	0.90
Indigo	CCR	1	1	1	1	1	1	1
	BCC-I	1	1	1	1	1	1	1
	BCC-O	1	1	1	1	1	1	1
Jet Airways	CCR	0.94	1	0.87	0.77	–	–	–
	BCC-I	1	1	1	0.77	–	–	–
	BCC-O	1	1	1	0.95	–	–	–
Jet Lite	CCR	0.49	0.41	0.29	0.08	–	–	–
	BCC-I	0.58	0.46	0.34	0.38	–	–	–
	BCC-O	0.97	0.88	0.86	0.91	–	–	–
Spice Jet	CCR	0.99	1	0.97	0.67	0.93	1	1
	BCC-I	1	1	1	1	1	1	1
	BCC-O	1	1	1	1	1	1	1
Vistara	CCR	0.54	0.51	0.51	0.15	0.31	0.52	0.33
	BCC-I	0.68	0.56	0.51	0.59	0.52	0.52	0.35
	BCC-O	0.76	0.77	0.85	0.92	0.92	0.90	0.86
True Jet	CCR	–	0.79	1	0.18	0.4	0.91	0.56
	BCC-I	–	1	1	1	1	1	0.76
	BCC-O	–	1	1	1	1	1	0.72
Zoom Air	CCR	–	–	1	1	–	–	
	BCC-I	–	–	1	1	–	–	–
	BCC-O	–	–	1	1	–	–	–
Star Air	CCR	–	–	–	–	0.51	1	1
	BCC-I	–	–	–	–	1	1	1
	BCC-O	–	–	–	–	1	1	1
Air Deccan	CCR	–	–	–	0.25	–	–	–
	BCC-I	–	–	–	1	–	–	–
	BCC-O	–	–	–	1	–	–	–

(continued)

Table 1.3 (continued)

		2014–2015	2015–2016	2016–2017	2017–2018	2018–2019	2019–2020	2020–2021
Air Odisha	CCR	–	–	–	1	–	–	–
	BCC-I	–	–	–	1	–	–	–
	BCC-O	–	–	–	1	–	–	–
Air Heritage	CCR	–	–	–	–	1	–	–
	BCC-I	–	–	–	–	1	–	–
	BCC-O	–	–	–	–	1	–	–

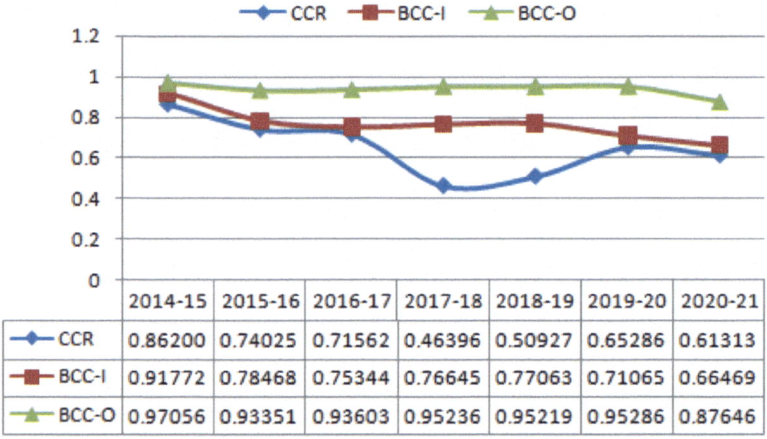

Fig. 1.1 Average efficiency scores of operating airlines for 2014–2021 using conventional models

The chart in Fig. 1.1 represents the performance trend of the domestic passenger airlines operating in India for 2014–2021. Performance scores with respect to the BCC model are continuously decreasing over the period. Concerning the CCR model, it first reduces in the first half and then increases later with a decrease in 2020. The sector's operational performance has declined in the last six years. On the contrary, as per the annual reports of the aviation sector www.dgca.gov.in, in the last few years, the airline sector has seen a significant hike in passenger growth.

1.4.2 Efficiency Scores Using Window Analysis

Window analysis was employed by Charnes et al. (1984) for air force problems. The technique introduces more freedom to the performance analysis as it is a non-parametric technique to measure the efficiency of DMU_S with respect to its performances over time. Since the data set under study has only eight numbers of DMU_S, this approach multiplies the data points included in the study. Therefore, it is advantageous for the efficiency analysis.

Table 1.4 presents the window analysis results for the Indian passenger airlines from 2014–2021 using the CCR model. The airlines' operational performance over the seven years is studied for different window lengths. The window length equals the number of years considered for a window. The second column in Table 1.4 indicates the window lengths. If the window length is four years for 2014–2021, then there will be four windows of length 4 (2014–2018, 2015–2019, 2016–2020, 2017–2021). For the first window of length 4, the airline is treated as a different DMU for 2014–2015, 2015–2016, 2016–2017, and 2017–2018. Thus, when Air India is assessed for 2014–2015, its performance in 2015–2016, 2016–2017, and 2017–2018 are in the constraints along with the 4-length window performances of other airlines.

The first, second, third, and fourth window columns contain the average efficiency scores for the window of respective lengths. The last column of Table 1.4 includes the average efficiency score of the airline for the period 2014–2021. Indigo and Air India Express attain the highest and lowest efficiency scores.

Table 1.5 presents the window analysis results over the period 2014–2021 for the length of window four using the BCC model with input orientation. As mentioned in Sect. 1.2.3, using a small window length will make the results for DEA window analysis more reliable.

The horizontal rows against each airline represent the overlapping windows, and each presents the performance trend of the corresponding airline for four consecutive years. It depicts how efficiently the airline is performing over a while. The vertical efficiency scores in the overlapping windows indicate the stability of the airlines' performance scores. The last column indicates the overall average score. In contrast to the CCR window analysis result, Spice Jet is the best performer, and Air India and Air India Express are the lowest performers throughout the study.

The last row of Table 1.5 presents the airline sector's average operational performance score for the corresponding year. Scores show a downward trend for two years, a sudden upward trend for the third year, and a downward trend followed by an upward trend. Thus, it can be observed that

- Air India and Air India Express have shown similar performance behavior. For every window, the performance trend was downward except for the score of the year 2020–2021.
- Air Asia, Vistara, and Go Air were progressive till 2017–2018. Later on, their performance score has shown a downward trend.
- Spice Jet and Indigo performance scores are more stable across the windows and show an upward trend throughout the period.
- It can be seen that Alliance Air started very well and went smoothly till 2017–2018. But later on, the operational performance has declined.
- There was a sudden decrement in the operational performance in the year 2018–2019 for all the airlines, and later on, due to COVID, the efficiency scores of 2019–2020 and 2020–2021 can't depict the real performance of the airline sector.

Table 1.4 Average CCR-I window efficiency scores of airlines for the period 2014–2021

	Length of window	First window	Second window	Third window	Fourth window	Average per length	Average per period
Air India	7	0.47	–	–	–	0.47	
	6	0.49	0.42	–	–	0.46	
	5	0.55	0.44	0.43	–	0.47	
	4	0.64	0.49	0.46	0.43	0.51	0.48
Air India Express	7	0.40	–	–	–	0.40	
	6	0.43	0.39	–	–	0.41	
	5	0.45	0.40	0.44	–	0.43	
	4	0.47	0.41	0.53	0.44	0.46	0.43
Alliance Air	7	0.75	–	–	–	0.75	
	6	0.77	0.71	–	–	0.74	
	5	0.81	0.73	0.87	–	0.80	
	4	0.86	0.76	0.89	0.85	0.84	0.78
Air Asia	7	0.44	–	–	–	0.44	
	6	0.59	0.43	–	–	0.51	
	5	0.6	0.58	0.46	–	0.55	
	4	0.62	0.59	0.70	0.45	0.59	0.52
Go Air	7	0.54	–	–	–	0.54	
	6	0.61	0.56	–	–	0.59	
	5	0.64	0.61	0.61	–	0.62	
	4	0.66	0.64	0.73	0.64	0.67	0.61
Indigo	7	0.83	–	–	–	0.83	
	6	0.86	0.86	–	–	0.86	
	5	0.85	0.88	0.91	–	0.88	
	4	0.89	0.87	0.94	0.96	0.92	0.87
Spice Jet	7	0.76	–	–	–	0.76	
	6	0.90	0.81	–	–	0.86	
	5	0.88	0.95	0.83	–	0.89	
	4	0.87	0.94	0.97	0.88	0.92	0.86
Vistara	7	0.42	–	–	–	0.42	
	6	0.52	0.40	–	–	0.46	
	5	0.52	0.52	0.44	–	0.49	
	4	0.53	0.52	0.62	0.45	0.53	0.48

Table 1.5 Average BCC-I window efficiency scores of airlines for 2014–2021

	2014–2015	2015–2016	2016–2017	2017–2018	2018–2019	2019–2020	2020–2021	Average
Air India	0.99	0.63	0.55	0.55	–	–	–	
	–	0.57	0.49	0.48	0.47	–	–	
	–	–	0.54	0.49	0.45	0.41	–	
	–	–	–	0.45	0.41	0.40	0.48	0.52
Air India Express	0.67	0.60	0.48	0.46	–	–	–	
	–	0.60	0.49	0.46	0.47	–	–	
	–	–	0.58	0.58	0.53	0.48	–	
	–	–	–	0.45	0.45	0.45	0.59	0.52
Alliance Air	1.00	1.00	0.68	1.00	–	–	–	
	–	1.00	0.70	1.00	0.65	–	–	
	–	–	1.00	1.00	0.88	0.94	–	
	–	–	–	1.00	0.83	0.87	1.00	0.91
Air Asia	0.63	0.54	0.69	0.76	–	–	–	
	–	0.54	0.69	0.76	0.54	–	–	
	–	–	0.78	0.81	0.60	0.65	–	
	–	–	–	0.75	0.53	0.58	0.45	0.64
Go Air	0.70	0.68	0.72	0.78	–	–	–	
	–	0.68	0.72	0.78	0.71	–	–	
	–	–	0.77	0.81	0.72	0.66	–	
	–	–	–	0.78	0.71	0.66	1.00	0.74
Indigo	0.92	0.83	0.88	1.00	–	–	–	
	–	0.82	0.86	0.98	1.00	–	–	
	–	–	0.84	0.92	1.00	1.00	–	
	–	–	–	0.92	1.00	1.00	1.00	0.94
Spice Jet	0.69	1.00	0.94	1.00	–	–	–	
	–	1.00	0.94	1.00	1.00	–	–	
	–	–	0.94	1.00	1.00	1.00	–	
	–	–	–	1.00	1.00	1.00	1.00	0.97
Vistara	0.65	0.56	0.56	0.65	–	–	–	
	–	0.56	0.56	0.65	0.51	–	–	
	–	–	0.68	0.69	0.58	0.65	–	
	–	–	–	0.65	0.50	0.50	0.56	0.59
	0.78	0.73	0.71	0.77	0.69	0.70	0.76	

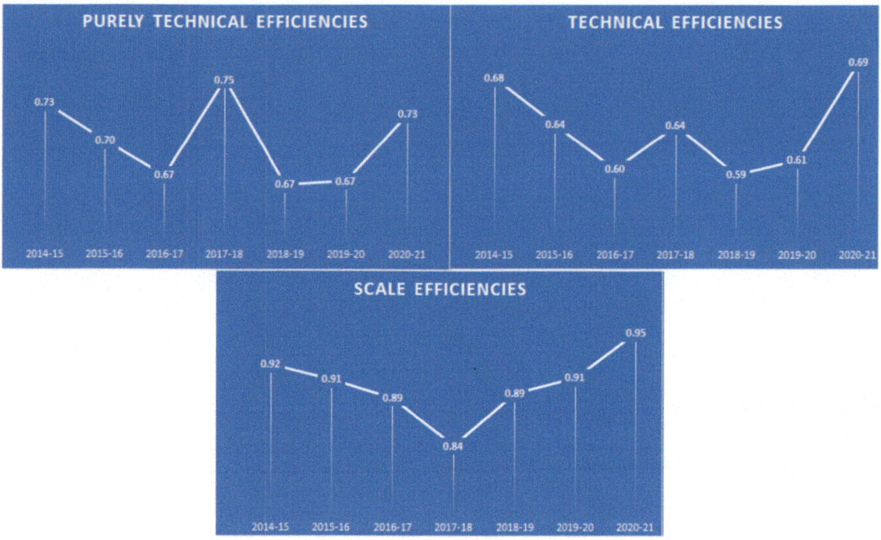

Fig. 1.2 Average conventional scores of airline sector for 2014–2021

Figure 1.2 displays the average TE, PTE, and SE scores for the airline sector over the period 2014–2021. The higher SE scores indicate that the fleet size and manpower need not be expanded. It can be observed that the low TE scores are due to the low PTE scores. Thus, there is a need to utilize the resources efficiently to generate more revenue.

1.4.3 Benchmarking and Proposed Projections

The section presents the required projection percentage for the input and output performance indicators for Air India, Air India Express, Alliance Air, Air Asia, Go Air, and Vistara Airlines. The percentage indicates the amount by which the resources should be reduced, and the revenue should be targeted so that the operational performance can be improvised.

Table 1.6 illustrates the minimum, maximum, and median values of the projection percentage of the input and output variables that the inefficient airline companies can target to attain the best efficiency score. The minimum/maximum/median projection % values presented in the table for the respective variable have been calculated by considering the minimum/maximum/median of all the % projections over the period 2014–2021 using the BCC model with both the input and output orientations.

The results obtained in Table 1.6 show that the airline companies can streamline their working performance by cutting the operating expenses per fleet size (I_1) in a range from 0.15% (Air Asia) to 69% (Vistara) and escalating the passenger load factor

Table 1.6 Proposed variable projections (%) for the period 2014–2021

S. No.	DMU	Variables	Minimum	Maximum	Median
1	Air India	I_1	–3.16	–63.36	–44.22
		I_2	–22.63	–66.56	–57.61
		O_1	0.13	10.62	9.98
		O_2	9.36	10.62	9.98
2	Air India Express	I_1	–0.75	–59.38	–37.74
		I_2	–46.64	–70.06	–59.15
		O_1	1.19	22.83	13.57
		O_2	20.13	118.62	84.33
3	Alliance Air	I_1	–25.35	–59.19	–43.75
		I_2	–32.31	–82.95	–58.68
		O_1	3.59	45.12	11.24
		O_2	8.9	606.51	330.68
4	Air Asia	I_1	–0.15	–64.52	–29.7
		I_2	–0.96	–69.06	–44.21
		O_1	3.09	19.81	8.93
		O_2	12.14	999.9	260.86
5	Go Air	I_1	–6.28	–37.68	–21.02
		I_2	–10.21	–57.46	–30.3
		O_1	3.34	10.75	5.64
		O_2	16.54	87.84	39.18
6	Vistara	I_1	–10.43	–68.48	–32.12
		I_2	–22.5	–76.11	–44.28
		O_1	0.57	31.31	13.88
		O_2	17.24	954.35	176.86

(O_1) in a range from 0.13% (Air India) to 45% (Alliance Air). It has been observed that the projection values are high for the expense per employee (I_2) and passenger revenue (O_2) variables. Airline companies can improvise the variables according to their working feasibility conditions and as per the given range of projections. Compared to private airlines, government airlines need to focus more on resource rationalization.

1.5 Conclusion

This paper applied the DEA window analysis technique to study the performance trend of the Indian passenger airlines in India for 2014–2021. Two inputs (operating expense per fleet size and expense per employee) and two outputs (passenger load factor and revenue generated from the sale of tickets) are employed to study the airlines' operational performance. CCR and BCC models are used for studying the TE, PTE, and SE scores over the period 2014–2021.

Window analysis DEA technique is used for the dynamic performance appraisal of the Indian airline sector during 2014–2021. The empirical investigation helps to understand the effect of variables under study on the operational performance of the airlines over a period of time.

In the future, research can be done to study the relationship between the variables, and sensitivity analysis can be done to study the effect on the operational efficiency of the variables. Decision-makers can benefit significantly from this empirical study and discussion of the results in order to optimize the operational efficiency of passenger airlines.

References

Al-Refaie A, Najdawi R, Al-Tahat M, Bata N (2015) Window analysis and malmquist index for assessing efficiency in a pharmaceutical industry. In: Proceedings of the world congress on engineering, vol 1. Newswood Limited (2015)

Asmild M, Paradi JC, Aggarwall V, Schaffnit C (2004) Combining dea window analysis with the malmquist index approach in a study of the canadian banking industry. J. Prod. Anal. 21:67–89

Banker RD, Charnes A, Cooper WW (1984) Some models for estimating technical and scale inefficiencies in data envelopment analysis. Manag Sci 30:1078–1092 (1984). https://doi.org/10.1287/mnsc.30.9.1078

Banker RD, Thrall RM (1992) Estimation of returns to scale using data envelopment analysis. Euro. J. Oper. Res. 62(1):74–84

Charnes A, Clark CT, Cooper WW, Golany B (1984) A developmental study of data envelopment analysis in measuring the efficiency of maintenance units in the u.s. air forces. Ann Oper Res 2:95–112 (1984). https://doi.org/10.1007/BF01874734

Charnes A, Cooper WW, Rhodes E (1978) Measuring the efficiency of decision making units. Euro J Oper Res 2:429–444 (1978). https://doi.org/10.1016/0377-2217(78)90138-8

Cooper WW, Seiford LM (2011) Handbook on data envelopment analysis

Farrell MJ (1957) The measurement of productive efficiency. https://www.jstor.org/stable/2343100

Joo SJ, Fowler KL (2014) Exploring comparative efficiency and determinants of efficiency for major world airlines. Benchmarking 21:675–687 (2014). https://doi.org/10.1108/BIJ-09-2012-0054

Kottas AT, Madas MA, Sahoo BK, Zhu J, Tone K, Klemen BM, Seiford LM, Zhu J (2018) Comparative efficiency analysis of major international airlines using data envelopment analysis: exploring effects of alliance membership and other operational efficiency determinants. J Air Transp Manag 70:1–17 (2018). https://doi.org/10.1016/j.jairtraman.2018.04.014

Miliotis PA (1992) Data envelopment analysis applied to electricity distribution districts. J Oper Res Soc 43 (1992). https://doi.org/10.1057/jors.1992.80

Saranga H, Nagpal R (2016) Drivers of operational efficiency and its impact on market performance in the indian airline industry. J. Air Transp. Manag. 53:165–176 (2016). https://doi.org/10.1016/j.jairtraman.2016.03.001

Seth B, Saxena P, Arora S (2023) Operational performance of indian passenger airlines using hierarchical categorical dea approach. Int J Syst Assur Eng Manag. https://doi.org/10.1007/s13198-022-01852-6

Chapter 2
Finite Buffer Reverse Balking Two Server Heterogeneous Service Queuing Model with Retention of Impatient Customers

Bhupender Kumar Som, Vivek Kumar Sharma, and Sunny Seth

Abstract In this paper, a heterogeneous two server finite capacity stochastic queuing model with modified buffer function of reverse balking, customer impatience and retention of impatient customers is developed. The model is studied in steady-state. The probabilities of zero customers in the system, n customers in the system and full capacity of the system are obtained iteratively. Expected size of the system, along with the expected queue length, expected rate of impatience and expected rate of retention are obtained. The probabilistic and performance measures thus obtained can be of immense use to the firms for framing efficient and sustainable strategies to manage business effectively in this uncertain business environment.

Keywords Queuing model · Reverse balking · Reneging · Retention · Finite buffer queue

2.1 Introduction and Literature Survey

The dimensions of business and services have evolved over a period of time but the fundamentals of business remained the same. One of such fundamentals is observing customers' behavior and planning sustainable business strategies around it. In this highly uncertain business environment, all the firms are competing against each other to sustain in the market. Understanding the Customers' behavior is one of the

B. K. Som
GNIOT Institute of Management Studies, Uttar Pradesh, Knowledge Park – II, Greater Noida 201310, India
e-mail: bksoam@live.com

V. K. Sharma
JaganNath University, Jaipur, Rajasthan 302022, India
e-mail: vivek.sharma@jagannathuniversity.org

S. Seth (✉)
Jagan Institute of Management Studies, Rohini, Sector -5, New Delhi 110085, India
e-mail: sunnyseth2005@gmail.com

R. Arora and S. Arora (eds.), *Optimization Techniques for Sustainable Environment Under Uncertainty*, Engineering Optimization: Methods and Applications,
https://doi.org/10.1007/978-981-96-0389-3_2

key components for the firms to stay ahead of their competitors. Customers look for a pleasant experience for buying any product or service, but due to supply and demand, they have to wait in queues (physically at checkout counters of stores or virtual processing time in case of online e-commerce platforms) which often turns out to be the most unpleasant part. If the waiting time of customers in queues is not managed effectively it results in longer queues and hence customers may leave the queue due to long waiting time. Such customers in this era of social media can post their unpleasant experiences about the firm, which can hamper the brand image and goodwill of the firm adversely. It can ultimately lead to a loss of business for that firm. So it becomes necessary for the firms to manage queues effectively by understanding customer behavior. Most of the queuing and cost-models are largely impacted by the behavior of the customers in the system. There are various types of customers' behavior studied by researchers in the past, e.g., Haight (1959) studied the queuing system with impatient customers. He termed these customers as reneged or impatient customers. He mentioned that every customer may wait in the queue up to a threshold value of time, after that, the customer may choose to leave the queue without completing the service. Further, the work is extended by Ancker and Gafarian (1963 and 1963a) by studying queues with balking and reneging. They quoted that there are customers who may not join the queue by observing the length of the queue in anticipation of a longer waiting time. These customers are termed as balking customers in the queuing system. The other important aspect of a queuing system is the way service is provided to the customers. The queues forming in front of systems can be served by a single server channel or a multiple-server channel. At times the servers vary in their rate of providing the service. These servers are termed as servers with varying service rates or simply heterogeneous servers.

Since then a lot of work on balking, reneging and heterogeneous service is carried out by researchers in different fields of applications. Jain et al. (2014) mentioned that there are situations in business when a customer gets attracted towards the system by observing the longer queue. The customers' tend to believe that longer queues in front of the system ensure better quality of service at the right cost. In such cases, it is a tradeoff between anticipation of longer waiting time and quality service and customers choose quality service at times. Hence longer queue in front of the system associates a higher probability of joining the system. This behavior is termed as reverse balking. Som and Seth (2017a) studied encouraged arrivals when firms introduce deals and discounts to encourage more and more customers to join the system and studied single server finite capacity model in steady-state and extended this model to heterogeneous two server case with reneging and retention with encouraged arrivals in (2017b). Som and Kumar (2018) studied a heterogeneous server queuing model with reverse balking. They presented a sensitivity analysis of the model in steady-state and derived various performance measures and extended it to multi-server case in Kumar and Som (2020). Kumar et al. (2020) also studied transient solution of heterogeneous two server model with balking and catastrophe. Som et al. (2020) have studied an infinite capacity reverse balking feedback queuing model with two heterogeneous servers, reneging and retention. In (2021) Saikia &

Choudhury have studied a modified limited buffer reverse balking function in steady-state for a single server queue. Som and Seth (2023) used this modified limited buffer reverse balking function in Markovian queuing model with feedback, reneging and retention of impatient customers in single server case and extended it to heterogeneous two servers queuing model case in this paper. Any firm encountering the contemporary issues of reverse balking, customer impatience adopting strategies for retention of impatient customers having two heterogeneous servers providing service to the customers can use the results of this model to understand the system performance measures and can devise sustainable strategies to manage the system effectively.

2.2 Mathematical Model Formulation

Finite buffer reverse balking two server heterogeneous service queuing model with reneging and retention of reneged customers is formulated under the following assumptions:

1. Poisson arrivals occur with exponentially distributed inter-arrival times with parameter $\frac{1}{\lambda}$.
2. Customers are served through two heterogeneous servers, with service rates, μ_1 and μ_2 respectively having exponentially distributed service times.
3. An arriving customer joins server 1 with probabilty π_1 and joins server 2 with probability $\pi_2 = 1 - \pi_1$.
4. An impatient customer may leave the queue after threshold time T with exponentially distributed parameter ξ.
5. A retention strategy may be adopted to retain a leaving impatient customer with probability q = (1 − p).
6. System has a finite capacity, N.
7. Customers are served on first come first serve (FCFS) basis.
8. A customer a may reverse balk (refuses to join the system) with probability $\left(1 - \frac{1}{\alpha - n\beta}\right)$ and may not reverse balk (joins the system) with probability $\left(\frac{1}{\alpha - n\beta}\right)$ where α and β are constants to be chosen depending on past data such that $\alpha - n\beta > 0$ for n = 0, 1, 2...N-1.

2.3 Steady-State Model

Model is formulated in Steady-State as follows:

$$0 = -\frac{\lambda}{\alpha} P_{00} + \mu_1 P_{10} + \mu_2 P_{01} \tag{2.1}$$

$$0 = -\left(\frac{\lambda}{\alpha-\beta} + \mu_1\right)P_{10} + \mu_2 P_{11} + \frac{\lambda}{\alpha}\pi_1 P_{00} \tag{2.2}$$

$$0 = -\left(\frac{\lambda}{\alpha-\beta} + \mu_2\right)P_{01} + \mu_1 P_{11} + \frac{\lambda}{\alpha}\pi_2 P_{00} \tag{2.3}$$

$$0 = -\left(\frac{\lambda}{\alpha-2\beta} + \mu_1 + \mu_2\right)P_2 + (\mu_1 + \mu_2 + \xi p)P_3 + \frac{\lambda}{\alpha-\beta}P_1 \tag{2.4}$$

$$0 = -\left(\frac{\lambda}{\alpha-n\beta} + \mu_1 + \mu_2 + (n-2)\xi p\right)P_n + (\mu_1 + \mu_2 + (n-1)\xi p)P_{n+1} + \frac{\lambda}{\alpha-(n-1)\beta}P_{n-1} \tag{2.5}$$

$$0 = \frac{\lambda}{\alpha-(N-1)\beta}P_{N-1} - (\mu_1 + \mu_2 + (N-2)\xi p)P_N \tag{2.6}$$

where P_{00} is the probability that the system is empty, P_{10} is the probability that server 1 is serving a customer and server 2 is idle, P_{01} is the probability that server 2 is serving a customer and server 1 is idle and P_n is the probability that there are n customers in the system.

Also $P_2 = P_{11}$ and $P_1 = P_{01} + P_{10}$.

2.4 Steady-State Solution

Solving Eq. (2.1) to (2.6), we obtain

$$P_{10} = \left(\frac{\left(\frac{\lambda}{\alpha-\beta}\right) + (\mu_1 + \mu_2)\pi_1}{\frac{2\lambda}{\alpha-\beta} + \mu_1 + \mu_2}\right)\frac{\lambda}{\alpha\mu_1}P_{00} \tag{2.7}$$

$$P_{01} = \left(\frac{\left(\frac{\lambda}{\alpha-\beta}\right) + (\mu_1 + \mu_2)\pi_2}{\frac{2\lambda}{\alpha-\beta} + \mu_1 + \mu_2}\right)\frac{\lambda}{\alpha\mu_2}P_{00} \tag{2.8}$$

Adding (2.7) and (2.8), we get

$$P_1 = \left(\frac{\left(\frac{\lambda}{\alpha-\beta}\right) + \pi_1\mu_2 + \pi_2\mu_1}{\frac{2\lambda}{\alpha-\beta} + \mu_1 + \mu_2}\right)\frac{\lambda}{\alpha}\frac{(\mu_1 + \mu_2)}{\mu_1\mu_2}P_{00} \tag{2.9}$$

Adding Eq. (2.2) and (2.3) and using (2.1) and (2.9), we get

$$P_2 = \left(\frac{\left(\frac{\lambda}{\alpha-\beta}\right) + \pi_1\mu_2 + \pi_2\mu_1}{\frac{2\lambda}{\alpha-\beta} + \mu_1 + \mu_2} \right) \frac{\lambda}{\alpha} \left(\frac{\lambda}{\alpha-\beta} \right) \frac{1}{\mu_1\mu_2} P_{00} \tag{2.10}$$

Using Eq. (2.5) and generalizing, we obtain

$$P_n = \left(\frac{\left(\frac{\lambda}{\alpha-\beta}\right) + \pi_1\mu_2 + \pi_2\mu_1}{\frac{2\lambda}{\alpha-\beta} + \mu_1 + \mu_2} \right) \frac{\lambda}{\alpha} \left(\frac{\lambda}{\alpha-\beta} \right) \frac{1}{\mu_1\mu_2} \prod_{k=3}^{n} \left(\frac{\left(\frac{\lambda}{\alpha-(k-1)\beta}\right)}{\mu_1 + \mu_2 + (k-2)\xi p} \right) P_{00} \tag{2.11}$$

For $3 \leq n \leq N-1$.

Using (2.6) and (2.11), we obtain

$$P_N = \left(\frac{\left(\frac{\lambda}{\alpha-\beta}\right) + \pi_1\mu_2 + \pi_2\mu_1}{\frac{2\lambda}{\alpha-\beta} + \mu_1 + \mu_2} \right) \frac{\lambda}{\alpha} \left(\frac{\lambda}{\alpha-\beta} \right) \frac{1}{\mu_1\mu_2} \prod_{k=3}^{N} \left(\frac{\left(\frac{\lambda}{\alpha-(k-1)\beta}\right)}{\mu_1 + \mu_2 + (k-2)\xi p} \right) P_{00} \tag{2.12}$$

Using condition of normality $\sum_{n=1}^{N} P_n = 1$, we get

$$P_0 = \left[\begin{array}{l} 1 + \left(\frac{\left(\frac{\lambda}{\alpha-\beta}\right) + \pi_1\mu_2 + \pi_2\mu_1}{\frac{2\lambda}{\alpha-\beta} + \mu_1 + \mu_2} \right) \frac{\lambda}{\alpha} \frac{(\mu_1+\mu_2)}{\mu_1\mu_2} + \left(\frac{\left(\frac{\lambda}{\alpha-\beta}\right) + \pi_1\mu_2 + \pi_2\mu_1}{\frac{2\lambda}{\alpha-\beta} + \mu_1 + \mu_2} \right) \frac{\lambda}{\alpha} \left(\frac{\lambda}{\alpha-\beta} \right) \frac{1}{\mu_1\mu_2} \\ + \sum_{n=3}^{N-1} \left(\frac{\left(\frac{\lambda}{\alpha-\beta}\right) + \pi_1\mu_2 + \pi_2\mu_1}{\frac{2\lambda}{\alpha-\beta} + \mu_1 + \mu_2} \right) \frac{\lambda}{\alpha} \left(\frac{\lambda}{\alpha-\beta} \right) \frac{1}{\mu_1\mu_2} \prod_{k=3}^{n} \left(\frac{\left(\frac{\lambda}{\alpha-(k-1)\beta}\right)}{\mu_1 + \mu_2 + (k-2)\xi p} \right) \\ + \left(\frac{\left(\frac{\lambda}{\alpha-\beta}\right) + \pi_1\mu_2 + \pi_2\mu_1}{\frac{2\lambda}{\alpha-\beta} + \mu_1 + \mu_2} \right) \frac{\lambda}{\alpha} \left(\frac{\lambda}{\alpha-\beta} \right) \frac{1}{\mu_1\mu_2} \prod_{k=3}^{N} \left(\frac{\left(\frac{\lambda}{\alpha-(k-1)\beta}\right)}{\mu_1 + \mu_2 + (k-2)\xi p} \right) \end{array} \right]^{-1} \tag{2.13}$$

2.5 Measures of Performance

Using the probabilistic measures obtained above, various measures of performance can be derived as follows:

1. Expected System Size (L_s):

$$L_s = \sum_{n=1}^{N} nP_n = P_1 + 2P_2 + \sum_{n=3}^{N-1} nP_n + NP_N \tag{2.14}$$

2. Expected queue length (L_q):

$$L_q = \sum\nolimits_{n=3}^{N} (n-2)P_n = \sum\nolimits_{n=3}^{N-1} (n-2)P_n + (N-2)P_N \quad (2.15)$$

3. Average rate of reneging (R_r):

$$R_r = \sum\nolimits_{n=3}^{N} (n-2)\xi p P_n = \sum\nolimits_{n=3}^{N-1} (n-2)\xi p P_n + (N-2)\xi p P_N \quad (2.16)$$

4. Average rate of retention (R_R):

$$R_R = \sum\nolimits_{n=3}^{N} (n-2)\xi q P_n = \sum\nolimits_{n=3}^{N-1} (n-2)\xi q P_n + (N-2)\xi q P_N \quad (2.17)$$

The probabilistic and performance measures obtained above can be translated in MS Excel to obtain simulated results for specific values of independent parameters which firms encounter at any point of time.

2.6 Numerical Illustration

Numerical illustration of the above model is presented below:

It can be seen from the above table that as arrival rate, λ increases, the expected system size, expected queue length both increases. Also with more number of customers in the system, the average rate of customer impatience also increases and thus average rate of retaining these customers also increases in the above model.

It can be seen that as service rate of a server increases, expected system size and expected queue length both decreases as desired. Also due to increased service rate and queue length getting reduced, reneging rate also decreases.

These simulated results can give a clear view to the firms regarding various performance measures and then firms can devise effective sustainable strategies to manage the system and avoid any chaotic situations due to excess load on servers or any other system uncertainties.

Table 2.1 Variation in $L_s, L_q,$ *and* R_r, R_R with respect to varying rate of arrivals we take, $N = 10, \alpha = 2, \beta = 0.1, \mu_1 = 2, \mu_2 = 3, \pi_1 = 0.4, \xi = 0.2, p = 0.4$

Average rate of arrival (λ)	Expected system size (Ls)	Expected queue length (L_q)	Average rate of reneging (R_r)	Average rate of retention (R_R)
2	0.43294	0.02094	0.00168	0.00251
2.5	0.56157	0.04326	0.00346	0.00519
3	0.70632	0.08014	0.00641	0.00962
3.5	0.87453	0.13853	0.01108	0.01662
4	1.07683	0.22850	0.01828	0.02742
4.5	1.32778	0.36407	0.02913	0.04369
5	1.64532	0.56290	0.04503	0.06755
5.5	2.04784	0.84391	0.06751	0.10127
6	2.54819	1.22179	0.09774	0.14661
6.5	3.14582	1.69948	0.13596	0.20394
7	3.82107	2.26207	0.18097	0.27145
7.5	4.53661	2.87704	0.23016	0.34525
8	5.24702	3.50243	0.28019	0.42029
8.5	5.91163	4.09881	0.32791	0.49186
9	6.50351	4.63831	0.37107	0.55660

Table 2.2 Variation in $L_s, L_q,$ *and* R_r with respect to varying rate of service we take, $N = 10, \alpha = 2, \beta = 0.1, \lambda = 5, \mu_2 = 2, \pi_1 = 0.4, \xi = 0.2, p = 0.4$

Average rate of service (μ_1)	Expected system size (Ls)	Expected queue length (L_q)	Average rate of reneging (R_r)
1.2	4.67293	2.98177	0.23854
1.4	4.08040	2.47277	0.19782
1.6	3.56045	2.03591	0.16287
1.8	3.11726	1.67218	0.13377
2	2.74665	1.37548	0.11004
2.2	2.44009	1.13641	0.09091
2.4	2.18761	0.94489	0.07559
2.6	1.97961	0.79161	0.06333
2.8	1.80764	0.66864	0.05349
3	1.66464	0.56951	0.04556
3.2	1.54490	0.48911	0.03913
3.4	1.44386	0.42344	0.03387
3.6	1.35791	0.36937	0.02955
3.8	1.28422	0.32453	0.02596
4	1.22054	0.28704	0.02296

2.7 Conclusion and Future Scope

In this paper, a queue with heterogeneous service, buffer function of reverse balking and retention of impatient customers is studied. Various probabilistic and performance measures are obtained and other measures of performance, e.g., waiting time distribution, expected waiting time in the queue, etc., can be derived with ease. The results obtained can help firms to understand the performance of the system well in advance and firms can plan sustainable strategies for effectively managing the business. The system can further be studied in a transient state and also optimization of the number of servers and cost model can be studied.

References

Ancker CJ, Gafarian AV (1963a) Some queuing problems with balking and reneging—II. Oper Res 11(6):928–937. https://doi.org/10.1287/opre.11.6.928

Ancker CJ, Gafarian AV (1963b) Some queuing problems with balking and reneging – I. Oper Res 11(1):88–100. https://doi.org/10.1287/opre.11.1.88

Armony M (2005) Dynamic routing in large-scale service systems with heterogeneous servers. Queueing Syst Theory Appl 51(3–4):287–329

Armony M, Mandelbaum A (2012) Routing and staffing in largescale service systems: the case of homogeneous impatient customers and heterogeneous servers. Oper Res 59(1):50–65

Efrosinin D, Sztrik J (2018) An algorithmic approach to analysing the reliability of a controllable unreliable queue with two heterogeneous servers. J Appl Probab 271(3):934–952

Haight FA (1959) Queueing with reneging. Metrika 2(1):186–197. https://doi.org/10.1007/bf0261 3734

Haji B, Ross MS (2015) A queueing loss model with heterogeneous skill based servers under idle time ordering policies. J Appl Probab 52(1):269–277

Jain M, Singh P (2001) M/M/m queue with balking reneging and additional servers. IJE Trans Basics 15(2):169–172

Kim HJ, Ahn SH, Righter R (2011) Managing queues with heterogeneous servers. J Appl Probab 48(2):435–452

Krishanamoorthy B (1963) On Poisson queues with heterogeneous servers. Oper Res 11:321–330

Kumar R, Som BK (2020) A multi-server queue with reverse balking and impatient customers. Pakistan J Stat 36(2):91–101

Kumar R, Sharma S, Soodan BS, Vijaya Laxmi P, Som BK (2020) Transient solution of a heterogeneous queuing system with balking and catastrophes. Reliab Theory Appl 15(1):42–50

Lin W, Kumar PR (1984) Optimal control of a queueing system with two heterogeneous servers. IEEE Trans Automat Control 29(8):696–703

Saikia G, Choudhury A (2021) A single server Markovian queuing system with limited buffer and reverse balking. Indep J Manag Product 12(7):1774–1784

Som BK (2019) A stochastic feedback queuing model with encouraged arrivals and retention of impatient customers. Springer Proc Bus Econ 2019:261–272

Som BK, Seth S (2017a) An M/M/1/N queuing system with encouraged arrivals. Global J Pure Appl Mathem 17:3443–3453

Som BK, Seth S (2017b) An M/M/2/N queuing system with encouraged arrivals, heterogenous service and retention of impatient customers. Adv Model Optimiz 19:97–104

Som BK, Kumar R (2018) A heterogeneous queuing system with reverse balking and reneging. J Ind Product Eng 35(1):1–5

Som BK, Seth S (2018) M/M/c/N queuing systems with encouraged arrivals, reneging, retention and Feedback customers. Yugoslav J Operat Res 28(3):333–344

Som BK, Sharma VK, Seth S (2020) An M/M/2 heterogeneous service Markovian feedback queuing model with reverse balking, reneging and retention of reneged customers. In Advances in Computing and Intelligent Systems: Proceedings of ICACM 2019 (pp. 291–296). Springer Singapore

Som BK, Seth S (2023) A finite buffer reverse balking feedback Markovian queuing system with reneging and retention of impatient customers. Pakistan J Stat 39(1):1–10

Chapter 3
Reckoning of Optimal Parameters of M/G/1 Retrial Model with Active and Passive Breakdowns Using Optimization Techniques

Radhika Agarwal and Divya Agarwal

Abstract Considering an M/G/1 queueing model with two types of breakdowns, i.e., active and passive breakdowns with immediate and delayed repairs respectively in this paper, global optimal values of salient measures to study the system are determined. The results obtained can be used in amplifying the serving standards of various sectors like banks, hospitals, telecommunication, etc. ANFIS (adaptive neuro-fuzzy inference system) is used to verify the influence of various boundary parameters on some performance measures. The cost function and the minimum cost for the system are obtained using two distinct optimization techniques, i.e., Firefly Algorithm (FA) and Cuckoo Search (CS). Furthermore, the minimum cost and minimum awaiting time of the system are estimated using multi objective genetic algorithm (MOGA). To ensure that the proposed solution is globally optimal, the minimization problem is unveiled as a convex programming problem.

Keywords Convexity · Firefly algorithm · Cuckoo search · MOGA · Multi-objective optimization · Unreliable server · Delayed repair

3.1 Introduction

Time and cost-efficient workflow can be developed in queueing systems to improve software congestion. Controllable queueing frameworks are expected to decide the ideal working strategy that limits the long-run cost. In a queueing model, the system is efficient if the total cost of the system is minimized (single-objective optimization). The developing anxiety of clients has progressed to the point where many

R. Agarwal · D. Agarwal (✉)
Amity Institute of Applied Sciences, Amity University, Sector 125, Noida, UP, India
e-mail: dagarwal1@amity.edu

R. Arora and S. Arora (eds.), *Optimization Techniques for Sustainable Environment Under Uncertainty*, Engineering Optimization: Methods and Applications,
https://doi.org/10.1007/978-981-96-0389-3_3

individuals are reluctant to wait in queues for the briefest span. Thus, halting time in long queues at banks, checkout counters and ticket counters is a service attribute that has become important to be considered while optimizing a problem. Thus, along with optimization of cost, we have optimized the halting time in queues and hence have got the efficient solution for the bi-objective problem using Multi-Objective Genetic Algorithm (MOGA).

Developing the cost-efficient model using optimization tools without discussing the convexity of the cost function is seen in most of the papers. Checking convexity in the queueing system is of substantial importance as it guarantees the presence of global optima. It is one of the most significant results of the optimization model. Taking inspiration from here, discussing the convexity of the cost function of the model discussed by Gao et al. (2020). From the convexity results, we confirm that the optimal retrial and repair rates obtained using FA and CS are global minima.

We come across many queueing situations where the server is inconsistent. Here considering a $M/G/1$ retrial system with active and passive breakdowns with delayed repairs during passive breakdown. *Active breakdowns*, as the name suggests, are the breakdowns when the system is actively working (i.e., busy) while *Passive breakdowns* occur if the system is inactive (i.e., in idle state). The model can be implemented in hospitals, network switching systems, cash counter banks, and supermarkets where service provider depends on computer systems for the service that is subject to breakdown. Then the main focus shifts to amendment of the system to avoid building up of long queues. All these queueing problems provide a perception to the model formulators to develop a model that can reduce the halting time of the clients.

3.2 Review of the Literature Work

Numerous scholars have investigated stochastic models. By adding a second optional service, breakdowns, and delayed repairs, where breakdowns happen at random while serving clients, Choudhury and Tadj (2009) extended the traditional $M/G/1$ model. They also assessed the busy and halting times as well as reliability indices. Seeing the recent developments, Singh et al. (2020) and Upadhyaya (2020) elaborated the $M^X/G/1$ queueing model with retrials taking into account customers' impatience with their feedback and slower (delayed) repairs. The earlier one analyzed reliability and cost optimization along with maximum entropy analysis for their model. While the latter is considered a well-known soft computing technique, ANFIS helped validate the analytical results of the model studied, and then a cost-effective system was built. Recently, Jain and Dhibar Jain and Dhibar (2023) dealt with the queueing analysis and strategic behavior of primitive users (PUs) and ordinary users (OUs) in cognitive radio systems (CRS). They have employed ANFIS for the numerical simulation and sensitivity analysis of various system metrics.

Numerous works show the optimization of the cost function but nowadays one can see multi-objective optimization where more than one objective function is optimized simultaneously so that one can obtain an efficient (Pareto optimal) solution to the problem. Verma (1986) first presented the idea of multi-objective optimization in queueing systems in the paper. Using Kahag et al. (2019) carried out a bi-objective optimization of the $M/M/m$ model to manage traffic congestion during hub allocation. Most recently, Wu and Yang (2021) considered a two phase mixed service model. They used Canonical PSO to do the single objective optimization and then constructed a bi-objective model to optimize expected cost and halting time simultaneously.

In literature, most of the papers based on obtaining the optimal cost do not take into account the sufficiency conditions of optimality which talks about the concept of convexity. It is an important characteristic that an objective function in any minimization problem holds to guarantee the optimality of the solution. A handful of authors have contributed in this direction in queueing papers. (Zhang et al., 1997) dealt with a $M/G/1$ model with two different vacations. They established the average cost function with the best threshold rules/policies. Later Zhang (2006) referred to the same paper and further evaluated the convexity of the average cost function. Gao et al. (2009) have discussed an unreliable retrial queue and a normal queue having unlimited capacity where the clients join the virtual track if the server breaks. They have done the convexity analysis and obtained the optimal retrial and repair rates.

Motivated by the pertinence of the above work, we have done a cost and convexity analysis of the $M/G/1$ retrial queueing model under active and passive breakdown services and delayed repairs studied by Zhang et al. (2020). They have analyzed the reliability and sojourn time for the system whereas the bi-objective optimization of the system is done using MOGA. The main focus remained on checking the convexity of the cost function to confirm that the optimal obtained from FA and CS is the global optimal. Further, for the bi-objective optimization, the Pareto optimal solution is determined. The developed model is applicable in computer and mobile networking systems, call centers, manufacturing and operating systems, and workplaces wherever these systems are used.

The rest of the paper is composed as follows. Section 3.3 consists of the basic definitions and preliminaries. Section 3.4 incorporates the basic model with the notations and analysis of the performance measures of the model. ANFIS validation and several factors' influence on queue length can be seen in Sect. 3.5. Section 3.6 gives the single and bi-objective optimization of the model. The paper's concluding remarks are included in Sect. 3.7.

3.3 Preliminaries

The purpose of this section is to consider some basic notations and definitions that are to be used throughout the chapter. Let for $\bar{x}, \bar{y} \in R^n$,

$$\bar{x} \leqq \bar{y} \implies \bar{x}_j \leq \bar{y}_j \ \forall j$$
$$\bar{x} \leq \bar{y} \implies \bar{x}_j \leq \bar{y}_j, \ \bar{x}_j \neq \bar{y}_j \ \forall j$$
$$\bar{x} < \bar{y} \implies \bar{x}_j < \bar{y}_j \ \forall j$$

Assume an unconstrained problem of optimization as

$$\text{(PP)} \qquad \text{Minimize} \quad \tau(y) = (\tau_1(y), \tau_2(y), \ldots \tau_u(y))$$

where Y is an open subset of R^n and $\tau : Y \to R^u$ are differentiable on Y. The problem (PP) is scalar if $u = 1$ else it is a multi-objective optimization problem.

Definition 1 For any function $\tau : R^n \to R$ is convex if $\forall\ \tilde{y} \in R^n$ we have $\tau[ky + (1-k)\tilde{y}] \leqq k\tau(y) + (1-k)\tau(\tilde{y})$ true $\forall\ k$ where $0 \leqq k \leqq 1$.

Definition 2 The Hessian matrix M is a square $n \times n$ matrix structured as follows if all partial derivatives of τ exist over the domain and are continuous

$$M = \begin{bmatrix} \frac{\partial^2 \tau}{\partial {x_1}^2} & \frac{\partial^2 \tau}{\partial x_1 \partial x_2} & \cdots & \frac{\partial^2 \tau}{\partial x_1 \partial x_n} \\ \frac{\partial^2 \tau}{\partial x_2 \partial x_1} & \frac{\partial^2 \tau}{\partial {x_2}^2} & \cdots & \frac{\partial^2 \tau}{\partial x_2 \partial x_n} \\ \vdots & \vdots & \ddots & \vdots \\ \frac{\partial^2 \tau}{\partial x_n \partial x_1} & \frac{\partial^2 \tau}{\partial x_n \partial x_2} & \cdots & \frac{\partial^2 \tau}{\partial {x_n}^2} \end{bmatrix}$$

Assume S to be a nonempty open convex set in R^n, and let $\tau : R^n \to R$ be twice differentiable on S. Then τ is said to be convex iff the Hessian matrix is positive semidefinite at each point in S. For more details refer to Bazaraa et al. (2005).

Definition 3 Global Optimal Point: If in MP $k = 1$, any point $\tilde{y} \in R^n$ is an optimal point if $\exists$ no $y \in R^n$ s.t. $\tau(y) < \tau(\tilde{y})$.

Definition 4 Global Pareto Optimal Point: A point $\tilde{y}$ is a global efficient/Pareto optimal solution of MP if $\exists$ no $y \in R^n$ s.t. $\tau(y) \leq \tau(\tilde{y})$.

A local Pareto solution is not always global Pareto optimal whereas the converse is always true. It holds only for a convex multi-objective programming problem. Also, if all the objectives and feasible regions are convex then the multi-objective optimization problem is convex. For more details refer to Miettinen (1999).

3.4 Representation of the Model

This segment describes the mathematical model developed by Gao et al. (2020) in Sect. 3.4.1 and states the performance measures and other important queueing measures such as system size, and halting time in Sect. 3.4.2.

3.4.1 Description of the Model

Considering an erratic retrial queue with two discrete active and passive breakdowns and postponed repairs as a result of the latter. The model's fundamental presumptions are:

- *Service policy*: If the server is vacant, the entering client gets the service right away. In case the server is inoperative or occupied then the client will have to retrial for service after some time.
- *Retrial policy*: In case the server is inoperative or occupied then the client will have to retrial for service following FCFS rule from a virtual orbit.
- *Breakdowns and repairs*: In occupied and unoccupied states, the service provider may face passive and active breakdowns respectively. The server is repaired immediately in case of active faults but in passive faults, the server stays unnoticed in the faults state and thus stays inoperative until a client enters and only then it gets its delayed repair. For both cases, the service is completed once the server is prepared to serve.

The following notations are used to describe the model:

λ, α	: Poisson arrival rate and exponential retrial rate of the customers.
δ, θ	: Breakdown rate for passive and active faults respectively (exponential distributed)
B, R, G	: Times to serve, repair of active and passive faults respectively (generally distributed)
$B(x), R(x), G(x)$	: Times to serve, repair of active and passive faults cumulative density function (cdf)
$b(x), r(x), g(x)$	: Times to serve, repair of active and passive faults probability generating function (pgf)
β_i, ν_i, μ_i	: Times to serve, repair of active and passive fault'si^{th} moment
$\beta(x), \nu(x), \mu(x)$	: Times to serve, repair of active and passive faults conditional complementary rates

3.4.2 Performance Indices

Under the stability condition $\rho + \frac{\delta\rho_1}{\lambda+\alpha+\delta} < \frac{\alpha}{\lambda+\alpha}$ and using the supplementary variable technique, Gao et al. (2020) obtained the steady state distribution along with the pgf of system size. Also, they have obtained the marginal probabilities and the following important performance indices as follows:

- The certainty Q_a that the service provider is unoccupied is

$$Q_a = Q_0(1) = \frac{\lambda[\delta + (\alpha + \lambda)(1 - \rho)]}{(\delta + \lambda)(\alpha + \lambda + \delta) + \delta\rho_1(\alpha + \lambda)}$$

- The likelihood Q_b that the server is busy (occupied) is

$$Q_b = Q_1(1) = \frac{\lambda(\alpha + \lambda) + \delta((\alpha + \lambda)\rho_1 + \delta + \alpha + 2\lambda)}{(\delta + \lambda)(\alpha + \lambda + \delta) + \delta\rho_1(\alpha + \lambda)}\lambda\beta_1$$

- The certainty Q_c that the server is under repair of active faults is

$$Q_c = Q_2(1) = \theta \nu_1 Q_1$$

- Q_d is the likelihood that the server is experiencing a delay in repair given as

$$Q_d = Q_3(1) = \frac{(\alpha + \lambda + \delta)(1 - \rho) - \lambda\rho - \delta\rho_1}{(\delta + \lambda)(\alpha + \lambda + \delta) + \delta\rho_1(\alpha + \lambda)}\delta$$

- The likelihood Q_e that the server is under repair of passive faults is

$$Q_e = Q_4(1) = \frac{\delta + (\alpha + \lambda)(1 - \rho)}{(\delta + \lambda)(\alpha + \lambda + \delta) + \delta\rho_1(\alpha + \lambda)}\delta\rho_1$$

The mean orbit (L_0) and system size (L_1) are

$$\begin{aligned} L_0 &= \frac{N''(1)}{2N'(1)} - \frac{D''(1)}{2D'(1)} \\ L_1 &= L_0 + Q_b + Q_c + Q_e \end{aligned} \tag{3.1}$$

where

$$\begin{aligned} N &= (z-1)(\alpha+\delta+\lambda)(\delta+\lambda) + \delta(\alpha+\lambda)(C(z) - H(z)) \\ N'(1) &= (\alpha+\delta+\lambda)(\delta+\lambda) + \delta(\alpha+\lambda)\rho_1 \\ N''(1) &= 2\delta\lambda\rho_1 + \delta(\alpha+\lambda)(\lambda^2\mu_2 + 2\rho\rho_1) \\ D'(1) &= (\alpha+\delta+\lambda)(\alpha+\lambda)\left[\frac{\alpha}{\alpha+\lambda} - \rho - \frac{\delta\rho_1}{\alpha+\lambda+\delta}\right] \\ D''(1) &= -(\alpha+\lambda)\left[2\lambda\rho + (\alpha+\lambda)(\lambda^2\beta_2^*)\right] \\ &\quad - \delta\left[2\lambda(\rho+\rho_1) + (\alpha+\lambda)(\lambda^2\mu_2 + 2\rho\rho_1 + \lambda^2\beta_2^*)\right] \end{aligned}$$

The halting time in the orbit (W_0) and the halting time in the system (W_1) are

$$W_0 = \frac{L_0}{\lambda} \text{ and } W_1 = \frac{L_1}{\lambda} \tag{3.2}$$

3.5 ANFIS Validation of Numerical Results

To carry out the numerical results, considering that repair and service time follow the Erlang distribution. The time to repair the passive fault is of order 5 with mean $\mu_1 = 1/\mu$ and $\mu_2 = 1/\mu^2$. The repair time of active failure is of order 2 with mean $\nu_1 = 1/\nu$ and $\nu_2 = 1/\nu^2$. The service times are of order 2 with mean $\beta_1 = 1/s$ and $\beta_2 = 1/s^2$. The retrial rate and passive and active breakdown rates follow an exponential distribution. The following parametric values are as follows: $\lambda = 0.1,\ \alpha = 3,\ \delta = 0.5,\ \theta = 0.5,\ \mu = 2,\ \nu = 4,\ s = 0.85$.

A powerful approach for identifying noteworthy results applicable in typical crowding conditions is ANFIS (adaptive network fuzzy inference system), a soft computing technique. It helps in perceiving estimated solutions for the measures whose exact outcomes are difficult to find accurately. Treating the parameters (α and s) in the ANFIS approach as linguistic variables that are run for three epochs each. Less, moderate, and high are specified as the linguistic values for each parameter. Presenting each one of the linguistic variables using the Gaussian membership function as shown in Figs. 3.1b and 3.2b for α and s, respectively. In Figs. 3.1a and 3.2a it is clear that the mathematical results coincide with that of neuro-fuzzy results for the model obtained by employing the ANFIS technique. Using this, the aim is to check the effect on mean system size (L_s) by changing α and s. It can be observed that L_1 rises exponentially with increasing retrials and slopes down logarithmically with rising service rate.

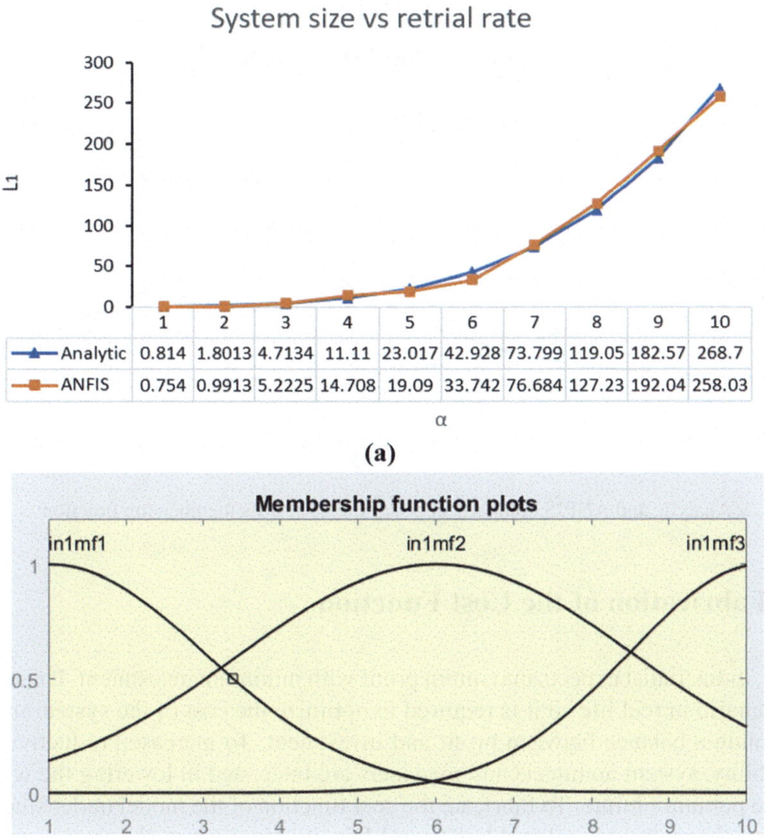

α	1	2	3	4	5	6	7	8	9	10
Analytic	0.814	1.8013	4.7134	11.11	23.017	42.928	73.799	119.05	182.57	268.7
ANFIS	0.754	0.9913	5.2225	14.708	19.09	33.742	76.684	127.23	192.04	258.03

Fig. 3.1 **a** Analytic and ANFIS comparison of L_1 versus α; **b** Its membership function

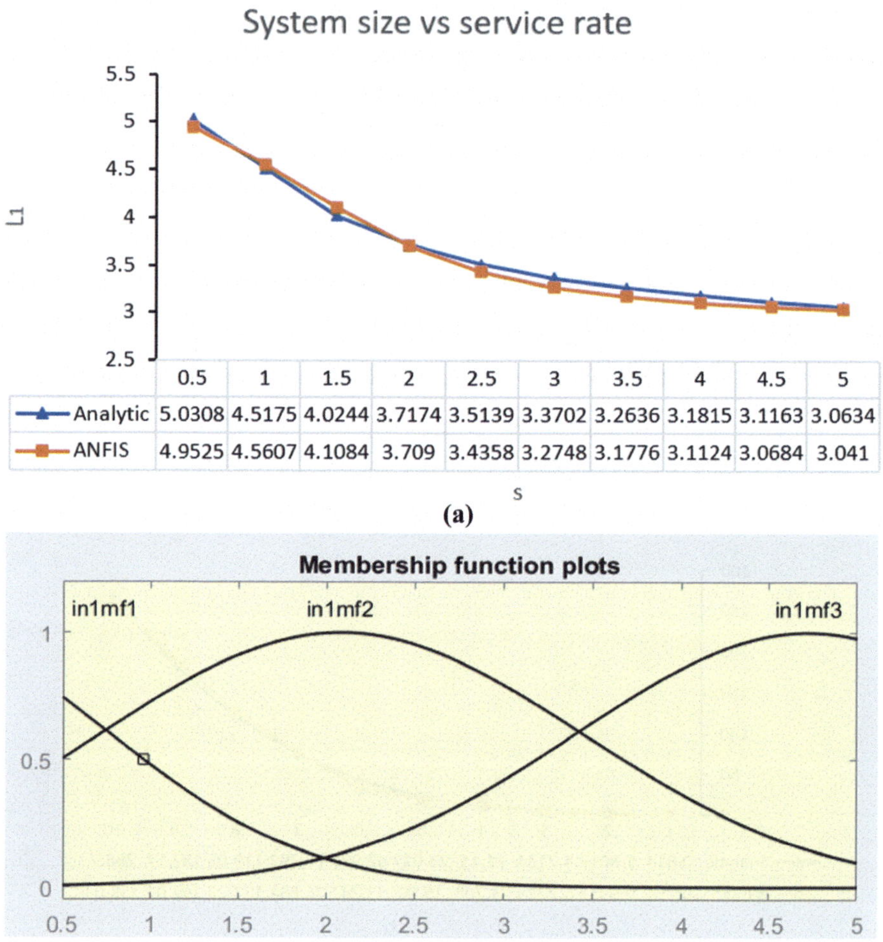

	0.5	1	1.5	2	2.5	3	3.5	4	4.5	5
Analytic	5.0308	4.5175	4.0244	3.7174	3.5139	3.3702	3.2636	3.1815	3.1163	3.0634
ANFIS	4.9525	4.5607	4.1084	3.709	3.4358	3.2748	3.1776	3.1124	3.0684	3.041

(a)

(b)

Fig. 3.2 **a** Analytic and ANFIS comparison of L_1 versus s; **b** Its membership function

3.6 Fabrication of the Cost Function

Mostly, industrialist expects maximum profit with minimum investment. It can never be a scenario in real life so it is required to optimize the cost of the system in order to maintain a balance between profit and investment. To increase productivity and profitability, system architects and managers are interested in lowering the working expense per unit of time. To fabricate the cost function of the model under study one can relate the mathematical problem to real-life situations where the service provider is the computing system and arrivals are the clients. In Gao et al. (2020), a detailed analysis of the system's cycle can be seen, from where employing the reasoning of a process of alternate renewal, getting

$$E[\theta] = \frac{1}{(\lambda + \delta)Q_{0,0}}$$

where

$$Q_{0,0} = \frac{\lambda}{\alpha}\left\{\frac{BC}{AB + \delta C\rho_1}\right\}\left(\frac{\alpha}{C} - \rho - \frac{\delta}{B}\rho_1\right)$$

$$A = \lambda + \delta, \quad B = \alpha + \lambda + \delta \quad and \quad C = \alpha + \lambda$$

Now constructing the total operating cost function (per unit time) of the queueing model which is defined as

$$TC = C_h L_1 + C_i Q_a + C_b Q_b + C_a Q_c + C_d Q_d + C_p Q_e + C_s \frac{1}{E[\theta]} \tag{3.3}$$

where
C_h : Cost for holding each client
C_i : Per unit time cost of computer system being unoccupied
C_b : Per unit time cost of computer system being occupied
C_a : Per unit time cost of computer system under repair for an active breakdown
C_d : Per unit time cost of computer system under delayed period
C_p : Per unit time cost of computer system under repair for a passive breakdown
C_s : Fixed set-up cost for the busy cycle.
The fundamental presumptions for important parameters for optimization are as follows:: $\lambda = 0.1,\ \alpha = 3,\ \delta = 0.5,\ \theta = 0.5,\ \mu = 2,\ s = 0.85,\ C_h = \$10,\ C_i = \$15,\ C_b = \$12,\ C_a = \$7,\ C_d = \$6,\ C_p = \$8,\ C_s = \$5.$

3.6.1 *Optimization of Single Objective Function*

In this subplot, our target is to cut down the system cost TC to perceive the optimal repair and retrial rates. The formula given in (3.3) is a nonlinear function of two continuous decision variables α and v. So the minimization problem can be considered as follows:

$$(CP) \qquad TC(\alpha^*, v^*) = \min TC(\alpha, v) \tag{3.4}$$

where $D : \{(\alpha, v) : \alpha, v \in (0.1, 8)\} \subset R^2$.

Clearly, D is a convex feasible set. One cannot minimize using conventional slope-based methods since the cost function comprises two continuous decision variables so aiming to deduce minimum cost using two different meta-heuristic techniques and compare the values. The two well-known techniques that are to be employed are the Firefly algorithm (FA) and Cuckoo Search (CS). FA was initially introduced by Yang in 2008which studies the flashing design and behavior of the normal fireflies while CS was developed by Yang and Deb together in 2009. They got inspiration from cuckoo's nature to lay their egg in other species' nests. According to the algorithm each egg depicts the solution while the cuckoo egg depicts a new solution. The

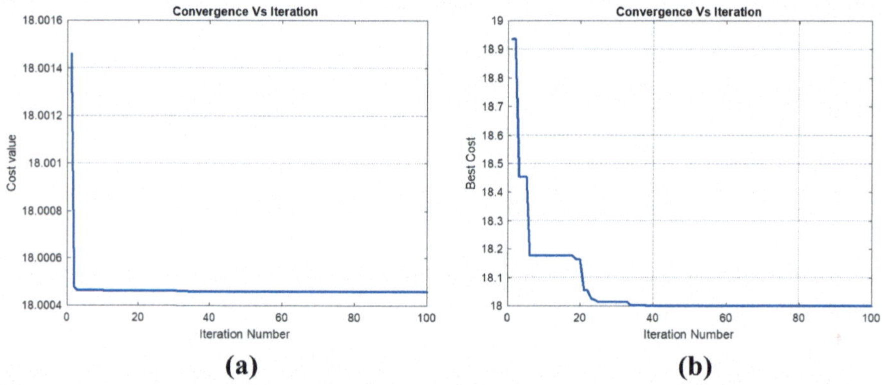

Fig. 3.3 Convergence versus iteration for firefly and Cuckoo search respectively in **a** and **b**

purpose is to get a better solution by replacing possibly finer results (cuckoos' eggs) with not-so-finer ones (already existing eggs).

Considering the MATLAB code for Firefly Algorithm, the system's total optimal cost obtained is $\mathbf{TC}(\alpha, \mathbf{v}) = \$\mathbf{18}$ for $(\alpha^*, \mathbf{v}^*) = (\mathbf{0.3922}, \mathbf{8})$ whereas the system's total optimal cost obtained using the MATLAB code of Cuckoo Search Algorithm is $\mathbf{TC}(\alpha, \mathbf{v}) = \$\mathbf{18}$ for $(\alpha^*, \mathbf{v}^*) = (\mathbf{0.3917}, \mathbf{8})$. Figure 3.3a and b show the convergence vs iteration graph for Firefly and Cuckoo Search algorithms respectively. One can see from the graphs that FA has a faster converging rate whereas CS converges a little more slowly but both give the same cost values with almost the same optimal values.

The cost function's convexity analysis is now discussed. Here, it can be demonstrated that the problem's objective function, (CP), is convex using the Hessian matrix test. Hence it is a convex programming problem (CPP). As a result, the ideal cost calculated above is a global minimum. Rewriting Eq. (3.3) to get

$$\begin{aligned} TC(\alpha, v) = {} & C_h(Q_i + Q_j + Q_b + Q_c + Q_e) + C_i Q_a + C_b Q_b \\ & + C_a Q_c + C_d Q_d + C_p Q_e + C_s(\lambda + \delta) Q_{0,0} \end{aligned} \tag{3.5}$$

The cost function's second-order partial derivatives are

$$
\begin{aligned}
\frac{\partial^2 TC}{\partial\alpha^2} =& C_h\left(\frac{\partial^2 Q_i}{\partial\alpha^2}+\frac{\partial^2 Q_j}{\partial\alpha^2}+\frac{\partial^2 Q_b}{\partial\alpha^2}+\frac{\partial^2 Q_c}{\partial\alpha^2}+\frac{\partial^2 Q_e}{\partial\alpha^2}\right)+C_i\frac{\partial^2 Q_a}{\partial\alpha^2}\\
&+C_b\frac{\partial^2 Q_b}{\partial\alpha^2}+C_a\frac{\partial^2 Q_c}{\partial\alpha^2}+C_d\frac{\partial^2 Q_d}{\partial\alpha^2}+C_p\frac{\partial^2 Q_e}{\partial\alpha^2}+C_s(\lambda+\delta)\frac{\partial^2 Q_{0,0}}{\partial\alpha^2}\\
\frac{\partial^2 TC}{\partial\alpha\partial v} =& C_h\left(\frac{\partial^2 Q_i}{\partial\alpha\partial v}+\frac{\partial^2 Q_j}{\partial\alpha\partial v}+\frac{\partial^2 Q_b}{\partial\alpha\partial v}+\frac{\partial^2 Q_c}{\partial\alpha\partial v}+\frac{\partial^2 Q_e}{\partial\alpha\partial v}\right)+C_i\frac{\partial^2 Q_a}{\partial\alpha\partial v}\\
&+C_b\frac{\partial^2 Q_b}{\partial\alpha\partial v}+C_a\frac{\partial^2 Q_c}{\partial\alpha\partial v}+C_d\frac{\partial^2 Q_d}{\partial\alpha\partial v}+C_p\frac{\partial^2 Q_e}{\partial\alpha\partial v}+C_s(\lambda+\delta)\frac{\partial^2 Q_{0,0}}{\partial\alpha\partial v}\\
\frac{\partial^2 TC}{\partial v^2} =& C_h\left(\frac{\partial^2 Q_i}{\partial v^2}+\frac{\partial^2 Q_j}{\partial v^2}+\frac{\partial^2 Q_b}{\partial v^2}+\frac{\partial^2 Q_c}{\partial v^2}+\frac{\partial^2 Q_e}{\partial v^2}\right)+C_i\frac{\partial^2 Q_a}{\partial v^2}\\
&+C_b\frac{\partial^2 Q_b}{\partial v^2}+C_a\frac{\partial^2 Q_c}{\partial v^2}+C_d\frac{\partial^2 Q_d}{\partial v^2}+C_p\frac{\partial^2 Q_e}{\partial v^2}+C_s(\lambda+\delta)\frac{\partial^2 Q_{0,0}}{\partial v^2}
\end{aligned}
\tag{3.6}
$$

where

$$
\begin{aligned}
\frac{\partial^2 Q_a}{\partial\alpha^2} &= \frac{2(\lambda+\delta(1+\rho_1))[\lambda\delta\rho A]}{[AB+\delta C\rho_1]^3}; \qquad \frac{\partial^2 Q_b}{\partial\alpha^2}=\frac{\partial^2 Q_c}{\partial\alpha^2}=0\\
\frac{\partial^2 Q_d}{\partial\alpha^2} &= \frac{2\delta(\lambda+\delta(1+\rho_1))[\delta^2(\rho+\rho_1)+\lambda\delta\rho]}{[AB+\delta C\rho_1]^3}; \qquad \frac{\partial^2 Q_e}{\partial\alpha^2}=\rho_1\frac{\partial^2 Q_d}{\partial\alpha^2}\\
\frac{\partial^2 Q_i}{\partial\alpha^2} &= \frac{[\lambda+\delta(1+\rho_1)][(\lambda)^2\mu_2+2\rho\rho_1]\delta A+2\lambda\delta\rho_1[\lambda+\delta(1+\rho_1)]^2}{[AB+\delta C\rho_1]^3}
\end{aligned}
$$

$$
\frac{\partial^2 Q_j}{\partial\alpha^2} = \frac{\begin{aligned}&[\alpha B-\rho BC-\delta C\rho_1]^2(\lambda)^2\beta_2^*-(1-\rho)[\alpha B-\rho BC-\delta C\rho_1]\\&\times\{C[2\lambda\rho+C(\lambda)^2\beta_2^*)]+\delta[2\lambda(\rho+\rho_1)+C((\lambda)^2\mu_2+2\rho\rho_1+(\lambda)^2\beta_2^*)]\}\\&-[\alpha B-\rho BC-\delta C\rho_1][\alpha+B-2\rho C-\delta(\rho+\rho_1)]\\&\times[2\lambda\rho+2C(\lambda)^2\beta_2^*+\delta((\lambda)^2\mu_2+2\rho\rho_1+(\lambda)^2\beta_2^*)]\\&+[\alpha+B-2\rho C-\delta(\rho+\rho_1)]^2\{C[2\lambda\rho+C(\lambda)^2\beta_2^*]\\&+\delta[2\lambda(\rho+\rho_1)+C((\lambda)^2\mu_2+2\rho\rho_1+(\lambda)^2\beta_2^*)]\}\end{aligned}}{\{BC[\frac{\alpha}{C}-\rho-\frac{\delta}{B}\rho_1]\}^3}
$$

$$
\begin{aligned}
\frac{\partial^2 Q_{0,0}}{\partial\alpha^2} = \lambda(\lambda+\delta)\Bigg\{&\left(\frac{BC}{\alpha(AB+\delta C\rho_1)}\right)\left(\frac{-2\lambda}{C^3}+\frac{-2\delta\rho_1}{B^3}\right)+\frac{Y\left(\frac{\alpha}{C}-\rho-\frac{\delta\rho_1}{B}\right)}{(\alpha^4)(AB+\delta C\rho_1)^4}\\
&+2\left(\frac{\alpha(AB+\delta C\rho_1)(2C+\delta)-BCX}{\alpha^2(AB+\delta C\rho_1)^2}\right)\left(\frac{\lambda}{C^2}+\frac{\delta\rho_1}{B^2}\right)\Bigg\}
\end{aligned}
$$

so that

$$X = [A(\alpha + B) + \delta(\alpha + C)\rho_1]$$
$$Y = \alpha^2(AB + \delta C\rho_1)^2[(AB + \delta C\rho_1)(2\lambda + 4\alpha + \delta) + (2\lambda + 2\alpha + \delta)X(\alpha - 1) - BC(2A + 2\delta\rho_1)] - [\alpha(AB + \delta C\rho_1)(2C + \delta) - BCX] \times [2\alpha(AB + \delta C\rho_1)^2 + 2\alpha^2(AB + \delta C\rho_1)X]$$

$$\frac{\partial^2 Q_a}{\partial\alpha\partial v} = \frac{2(\lambda)^2\beta_1\theta\,[(AB + \delta\rho_1 C) - C(\lambda + \delta(1+\rho))]}{v^2(AB + \delta\rho_1 C)^2}; \quad \frac{\partial^2 Q_b}{\partial\alpha\partial v} = 0$$
$$\frac{\partial^2 Q_c}{\partial\alpha\partial v} = \frac{-2\lambda\beta_1\theta(\lambda + \delta(1+\rho))\,[(AB + \delta\rho_1 C) - (\lambda C + \delta(C\rho_1 + \lambda + B))]}{v^2(AB + \delta\rho_1 C)^2}$$
$$\frac{\partial^2 Q_d}{\partial\alpha\partial v} = \frac{2\theta\lambda\delta\beta_1\,(AB + \delta\rho_1 C) - \delta C(\lambda + \delta(1+\rho))\left[\lambda\beta_1\left(\frac{2\theta}{v^2}\right)(B + \lambda)\right]}{v^2(AB + \delta\rho_1 C)^2}$$
$$\frac{\partial^2 Q_e}{\partial\alpha\partial v} = \frac{2\lambda\theta\beta_1\delta\rho_1\,[(AB + \delta\rho_1 C) - C(\lambda + \delta(1+\rho))]}{v^2(AB + \delta\rho_1 C)^2}$$

$$\frac{\partial^2 Q_{0,0}}{\partial\alpha\partial v} = (\lambda + \delta)2\theta\lambda^2\beta_1\left\{\frac{\alpha(AB + \delta\rho_1 C)(B + C) - BC[A(\alpha + B) + \delta\rho_1(\alpha + C)]}{v^2\alpha^2(AB + \delta\rho_1 C)^2}\right\}$$
$$\frac{\partial^2 Q_i}{\partial\alpha\partial v} = \frac{-2\lambda\theta\beta_1\delta\rho_1\,[(AB + \delta\rho_1 C) - C(\lambda + \delta(1+\rho))]}{v^2(AB + \delta\rho_1 C)^2}$$

$$\frac{\partial^2 Q_j}{\partial\alpha\partial v} = \frac{\left(4\lambda\theta\beta_1 BCv^2\right)\left\{\begin{array}{r} 2[(\alpha + B) - \rho(B + C) - \delta\rho_1]N' + 2BC\left(\frac{\alpha}{C} - \rho - \frac{\delta\rho_1}{B}\right)\frac{\partial N'}{\partial\alpha} \\ -\left(4\lambda\theta\beta_1(B + C)v^2\right)N - \frac{\partial N}{\partial\alpha}\left(4\lambda\theta\beta_1 BCv^2\right)\end{array}\right\} - 2[(\alpha + B) - \rho(B + C) - \delta\rho_1]\left\{2BC\left(\frac{\alpha}{C} - \rho - \frac{\delta\rho_1}{B}\right)N' - \left(4\lambda\theta\beta_1 BCv^2\right)N\right\}}{\left(2BC\left(\frac{\alpha}{C} - \rho - \frac{\delta\rho_1}{B}\right)\right)^3}$$

such that

$$N = C\left[2(\lambda)^2\beta_1\left(1 + \frac{2\theta}{v}\right) + C(\lambda)^2\left(\beta_2\left(1 + \frac{2\theta}{v}\right)^2 + \theta\beta_1\frac{2}{v^2}\right)\right] + \delta\left[2(\lambda)^2\beta_1\left(1 + \frac{2\theta}{v}\right) + 2\lambda\rho_1 + C\left(\begin{array}{c}(\lambda)^2\mu_2 + 2(\lambda)^2\beta_1\left(1 + \frac{2\theta}{v}\right)\rho_1 \\ +(\lambda)^2\left(\beta_2\left(1 + \frac{2\theta}{v}\right)^2 + \theta\beta_1\frac{2}{v^2}\right)\end{array}\right)\right]$$

$$N' = \lambda^2 BC\left[4\theta\beta_2\left(1+\frac{2\theta}{v}\right)\left(\frac{-1}{v^2}\right) - \frac{4\theta\beta_1}{v^3}\right] + 4\lambda\theta\beta_1\left(\frac{-1}{v^2}\right)(\lambda B + \delta\rho_1 C)$$

$$\frac{\partial N}{\partial \alpha} = 2\lambda\rho + 2C(\lambda)^2\beta_2^* + \delta((\lambda)^2\mu_2 + 2\rho\rho_1 + (\lambda)^2\beta_2^*)$$

$$\frac{\partial N'}{\partial \alpha} = (\lambda)^2\left[4\theta\beta_2\left(1+\frac{2\theta}{v}\right)\left(\frac{-1}{v^2}\right) - \frac{4\theta\beta_1}{v^3}\right](B+C) + 4\lambda\theta\beta_1\left(\frac{-1}{v^2}\right)(\lambda + \delta\rho_1)$$

$$\frac{\partial^2 Q_a}{\partial v^2} = \frac{-4(\lambda)^2\beta_1\theta C}{v^3(AB+\delta\rho_1 C)}; \quad \frac{\partial^2 Q_b}{\partial v^2} = 0; \quad \frac{\partial^2 Q_c}{\partial v^2} = \frac{4\lambda\beta_1\theta(\lambda C + \delta(C\rho_1 + \lambda + B))}{v^3\,(AB+\delta\rho_1 C)};$$

$$\frac{\partial^2 Q_d}{\partial v^2} = \frac{-4\lambda\theta\beta_1\delta(\lambda + B)}{v^3(AB+\delta\rho_1 C)}; \quad \frac{\partial^2 Q_e}{\partial v^2} = \frac{-4\lambda\delta\theta\beta_1 C\rho_1}{v^3\,(AB+\delta\rho_1 C)};$$

$$\frac{\partial^2 Q_{0,0}}{\partial v^2} = -4\theta\lambda\beta_1\frac{(\lambda+\delta)}{v^3}\frac{\lambda}{\alpha}\frac{BC}{(AB+\delta\rho_1 C)}; \quad \frac{\partial^2 Q_i}{\partial v^2} = \frac{4\lambda\theta\beta_1\delta\rho_1 C}{v^3\,(AB+\delta\rho_1 C)}$$

$$\frac{\partial^2 Q_j}{\partial v^2} = \frac{2BC\left(\frac{\alpha}{C} - \rho - \frac{\delta\rho_1}{B}\right)N_1'' - N_1 4BC\theta\lambda\beta_1\left(\frac{-2}{v^3}\right)}{\left[2BC\left(\frac{\alpha}{C} - \rho - \frac{\delta\rho_1}{B}\right)\right]^2}$$
$$- \frac{2\left[(2BC)^2\theta\lambda\beta_1\left(\frac{2}{v^2}\right)\left(\frac{\alpha}{C} - \rho - \frac{\delta\rho_1}{B}\right)N_1' - N_1\left(2BC\theta\lambda\beta_1\left(\frac{2}{v^2}\right)\right)^2\right]}{\left[2BC\left(\frac{\alpha}{C} - \rho - \frac{\delta\rho_1}{B}\right)\right]^3}$$

so that

$$N_1 = C\left[2(\lambda)^2\beta_1\left(1+\frac{2\theta}{v}\right) + C(\lambda)^2\left(\beta_2\left(1+\frac{2\theta}{v}\right)^2 + \theta\beta_1\left(\frac{2}{v^2}\right)\right)\right]$$
$$+\delta\left[2(\lambda)^2\beta_1\left(1+\frac{2\theta}{v}\right) + 2\lambda\rho_1 + C\left\{\begin{array}{c}(\lambda)^2\mu_2 + 2\lambda\beta_1\rho_1\left(1+\frac{2\theta}{v}\right)\\ +(\lambda)^2\left[\beta_2\left(1+\frac{2\theta}{v}\right)^2 + \theta\beta_1\left(\frac{2}{v^2}\right)\right]\end{array}\right\}\right]$$

$$N_1' = (\lambda)^2 BC\left[4\theta\beta_2\left(\frac{-1}{v^2}\right)\left(1+\frac{2\theta}{v}\right) + 2\theta\beta_1\left(\frac{-2}{v^3}\right)\right]$$
$$+ 4\theta(\lambda)^2\beta_1 B\left(\frac{-1}{v^2}\right) + 4\theta\lambda\beta_1\rho_1\delta C\left(\frac{-1}{v^2}\right)$$

$$N_1'' = (\lambda)^2 BC\left[4\theta\beta_2\left(\frac{2}{v^3}\right)\left(1+\frac{2\theta}{v}\right) + 4\theta\beta_1\left(\frac{3}{v^4}\right) + 8\theta^2\beta_2\left(\frac{-1}{v^2}\right)^2\right]$$
$$+ 4\theta(\lambda)^2\beta_1 B\left(\frac{2}{v^3}\right) + 4\theta(\lambda)^2\beta_1\rho_1\delta C\left(\frac{2}{v^3}\right)$$

It is obvious that all of TC's partial derivatives exist and are continuous over the domain, so the Hessian matrix of TC will be

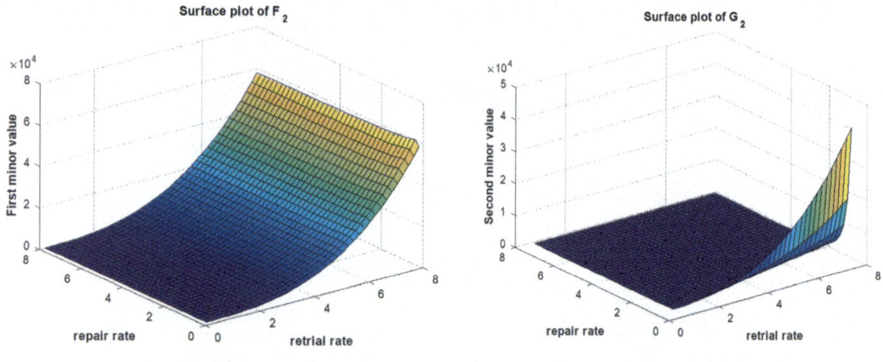

(a) **(b)**

Fig. 3.4 Surface plot of first minor and determinant of TC in **a** and **b** respectively

$$M_1 = \begin{bmatrix} \frac{\partial^2 TC}{\partial \alpha^2} & \frac{\partial^2 TC}{\partial \alpha \partial v} \\ \frac{\partial^2 TC}{\partial v \partial \alpha} & \frac{\partial^2 TC}{\partial v^2} \end{bmatrix}$$

Let's denote the determinant of the first and second principle minor as

$$F_1 = \left|\frac{\partial^2 TC}{\partial \alpha^2}\right|, \quad G_1 = \left|\left(\frac{\partial^2 TC}{\partial \alpha^2}\right)\left(\frac{\partial^2 TC}{\partial v^2}\right) - \left(\frac{\partial^2 TC}{\partial \alpha \partial v}^2\right)\right|$$

respectively. As clear from the 3D plots (Fig. 3.4a, b) one can see that $F_1 > 0$ and $G_1 \geqq 0$ so M_1 is a positive semidefinite matrix. Thus one can say that TC is indeed a convex function. As a result, the programming problem (CP) over D is convex. Thus, the system's obtained optimal cost, $\mathbf{TC}(\boldsymbol{\alpha}, \mathbf{v}) = \$\mathbf{18}$ is a global minimum.

3.6.2 Multi-Objective Optimization

A multi-objective optimization problem involves multiple conflicting objectives that need to be optimized simultaneously. For example, in engineering, one might want to minimize cost while maximizing performance, which are often conflicting goals. Multifarious inclusion is important to make accurate decisions, specifically if the objectives dispute within a limited source. Some studies on bi-objective models in queueing systems can be seen in Agarwal et al. (2022), Kahag et al. (2019), Mohammadi et al. (2014), Moghaddam et al. (2017), Wu and Yang (2021). In this subplot, constructing a bi-objective optimization problem where the expected cost function (TC) and the halting time in queues (W_1) are to be minimized at the same time over a domain D. The problem of bi-objective optimization can be considered as:

$$(MP) \qquad (TC^*, W_1{}^*) = \min(TC, W_1) \tag{3.7}$$

where $D : \{(\alpha, v) : \alpha, v \in (0.1, 8)\} \subset R^2$.

Multi-Objective Genetic Algorithm (MOGA) is inspired by the process of natural selection and is part of the broader family of evolutionary algorithms. It maintains a population of candidate solutions and applies genetic operators (such as mutation, crossover, and selection) to generate new candidate solutions. The key difference between MOGA and single-objective genetic algorithms is the way fitness is evaluated. In MOGA, fitness is determined based on multiple objectives simultaneously. This means MOGA aims to find a set of solutions that represent a trade-off among different objectives, rather than a single optimal solution. The set of solutions that represent the best trade-offs between conflicting objectives is known as the Pareto optimal frontier. It is to be noted that the Pareto optimal solutions from the Pareto frontier cannot improve both objective functions simultaneously. Thus an efficient solution is obtained such that the functions are balanced in a way that invests minimum cost to the system with minimum waiting time for the clients. Using MOGA one can see that for the global optimal rates $(\boldsymbol{\alpha}^*, \mathbf{v}^*) = (\mathbf{0.8}, \mathbf{7.8})$ to get $\mathbf{TC} = \$\mathbf{18.5}$ and $\mathbf{W_1} = \mathbf{6.9}$ mins to be the minimum cost and minimum waiting time for clients.

In optimization, a multi-objective problem is convex if all the objectives and the feasible region are convex. One can refer to Miettinen (1999) for further information. While determining whether (MP) is convex or not, both of the objective functions must be demonstrated to be convex to guarantee that the least value attained for (MP) is the global Pareto optimum value. As discussed in Sect. 3.6.1, the objective function TC is convex. To show the second objective function W_1 to be convex, rewriting Eq. (3.2) as

$$W_1(\alpha, v) = \frac{C_h}{\lambda}(Q_i + Q_j + Q_b + Q_c + Q_e) \tag{3.8}$$

The partial derivatives of the second order of Eq. (3.8) will be

$$\begin{aligned}
\frac{\partial^2 W_1}{\partial \alpha^2} &= \frac{C_h}{\lambda}\left(\frac{\partial^2 Q_i}{\partial \alpha^2} + \frac{\partial^2 Q_j}{\partial \alpha^2} + \frac{\partial^2 Q_b}{\partial \alpha^2} + \frac{\partial^2 Q_c}{\partial \alpha^2} + \frac{\partial^2 Q_e}{\partial \alpha^2}\right)\\
\frac{\partial^2 W_1}{\partial \alpha \partial v} &= \frac{C_h}{\lambda}\left(\frac{\partial^2 Q_i}{\partial \alpha \partial v} + \frac{\partial^2 Q_j}{\partial \alpha \partial v} + \frac{\partial^2 Q_b}{\partial \alpha \partial v} + \frac{\partial^2 Q_c}{\partial \alpha \partial v} + \frac{\partial^2 Q_e}{\partial \alpha \partial v}\right)\\
\frac{\partial^2 W_1}{\partial v^2} &= \frac{C_h}{\lambda}\left(\frac{\partial^2 Q_i}{\partial v^2} + \frac{\partial^2 Q_j}{\partial v^2} + \frac{\partial^2 Q_b}{\partial v^2} + \frac{\partial^2 Q_c}{\partial v^2} + \frac{\partial^2 Q_e}{\partial v^2}\right)
\end{aligned} \tag{3.9}$$

where Q_i, Q_j, Q_b, Q_c, Q_e are as defined in Sect. 3.6.1. The partial derivatives of the second order of W_1 exist and are continuous over the domain where the Hessian matrix is given as

$$M_2 = \begin{bmatrix} \frac{\partial^2 W_1}{\partial \alpha^2} & \frac{\partial^2 W_1}{\partial \alpha \partial v} \\ \frac{\partial^2 W_1}{\partial v \partial \alpha} & \frac{\partial^2 W_1}{\partial v^2} \end{bmatrix}$$

Think of determining the first principal minor and second principal minor as

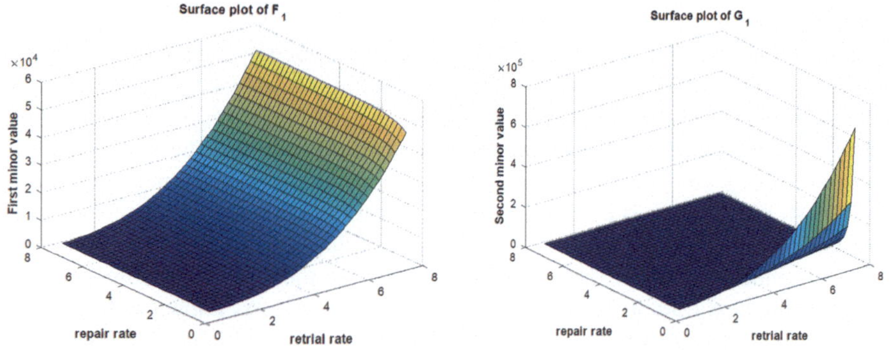

Fig. 3.5 Surface plot of first minor and determinant of TC in **a** and **b** respectively

$$F_1 = \left|\frac{\partial^2 W_1}{\partial \alpha^2}\right|, \quad G_1 = \left|\left(\frac{\partial^2 W_1}{\partial \alpha^2}\right)\left(\frac{\partial^2 W_1}{\partial \upsilon^2}\right) - \left(\frac{\partial^2 W_1}{\partial \alpha \partial \upsilon}^2\right)\right|$$

Figure 3.5a and b shows that $F_2 > 0$ and $G_2 \geqq 0$. So M_2 is a positive semidefinite matrix thus W_1 is also a convex function. Thus (MP) is a convex programming problem. Therefore, it can be clearly said that the Pareto optimal values $\mathbf{TC}^* = \$\mathbf{18.5}$ and $\mathbf{W_1}^* = \mathbf{6.9}$ are global best optimum values of an objective problem involving two variables (MP).

3.7 Conclusion

The developed model is applicable in all real-life scenarios where the server is erratic and can be acquired by system managers to obtain the minimized cost in less waiting time in the system. Cost-effective systems are required to tackle real-life situations constructively. Sensitivity analysis helps in getting the analytical results and the ANFIS technique helps managers looking to boost client happiness and service quality. The detailed convexity analysis enables us to ensure that the optimal points obtained using the Cuckoo search are not just optimal but are global optimal points. This makes our model reliable in the real economic world.

References

Agarwal R, Agarwal D, Upadhyaya S, Ahmad I (2022) Optimization of a stochastic model having erratic server with immediate or delayed repair. Ann Oper Res. https://doi.org/10.1007/s10479-022-04804-2

Bazaraa MS, Sherali HD, Shetty CM (2005) Nonlinear programming - theory and algorithms, 3rd edn. Wiley, New York

Choudhury G, Tadj L (2009) An M/G/1 queue with two phases of service subject to the server breakdown and delayed repair. Appl Math Model 33(6):2699–2709. https://doi.org/10.1016/j.apm.2008.08.006

Gao S, Zhang J, Wang X (2020) Analysis of a retrial queue with two-type breakdowns and delayed repairs. in IEEE Access 8:172428–172442. http://doi.org/10.1109/ACCESS.2020.3023191

Jain M, Dhibar S (2023) ANFIS and metaheuristic optimization for strategic joining policy with re-attempt and vacation. Math Comput Simul 211:57–84. https://doi.org/10.1016/j.matcom.2023.03.024

Kahag MR, Niaki STA, Seifbarghy M, Zabihi S (2019) Bi-objective optimization of multi-server intermodal hub-location allocation problem in congested systems: modeling and solution. J Ind Eng Int 15(2):221–248. https://doi.org/10.1007/s40092-018-0288-0

Mohammadi M, Dehbari S, Vahdani B (2014) Design of a bi-objective reliable healthcare network with finite capacity queue under service covering uncertainty. Transp Res Part E: Logist Transp Rev 72:15–41. https://doi.org/10.1016/j.tre.2014.10.001

Miettinen K (1999) Nonlinear multiobjective optimization. Kluwer Academic Publishers, Boston

Sherman NP, Kharoufeh JP, Abramson MP (2009) An M/G/1 retrial queue with unreliable server for streaming multimedia applications Probab Eng Inf Sci 23(2):281–304. https://doi.org/10.1017/S0269964809000175

Singh CJ, Kaur S, Jain M (2020) Unreliable server retrial G-queue with bulk arrival, optional additional service and delayed repair. Int J Oper Res 38(1):82–111. https://doi.org/10.1504/IJOR.2020.106362

Tavakkoli-Moghaddam R, Vazifeh-Noshafagh S, Taleizadeh AA, Hajipour V, Mahmoudi A (2017) Pricing and location decisions in multi-objective facility location problem with M/M/m/k queuing systems. Optim Eng 49(1):136–160. https://doi.org/10.1080/0305215X.2016.1163630

Upadhyaya S (2020) Investigating a general service retrial queue with damaging and licensed units: an application in local area networks. Opsearch 57:716–745. https://doi.org/10.1007/s12597-020-00440-1

Verma RK (1986) Multiobjective optimization of a queueing system. Optimization 17(1):103–115. https://doi.org/10.1080/02331938608843107

Wu CH, Yang DY (2021) Bi-objective optimization of a queueing model with two-phase heterogeneous service. Comput Oper Res 130(1):105230. https://doi.org/10.1016/j.cor.2021.105230

Yang XS (2008) Nature-inspired meta-heuristic algorithms. Luniver Press, Beckington

Yang XS, Deb S (2009) Cuckoo search via Lévy flights. In: In proceedings of world congress on nature & biologically inspired computing (NaBIC), IEEE, Coimbatore(India), vol 2009, pp 210–214

Zhang ZG, Vickson RG, Van Eenige MJA (1997) Optimal two-threshold policies in an M/G/1 queue with two vacation types. Perform Eval 29(1):63–80. https://doi.org/10.1016/S0166-5316(96)00005-3

Zhang ZG (2006) On the convexity of the two-threshold policy for an M/G/1 queue with vacations. Oper Res Lett 34(4):473–476. https://doi.org/10.1016/j.orl.2005.07.002

Chapter 4
Multi-objective Fuzzy Random Quadratic Transportation Problem

P. Anukokila and B. Radhakrishnan

Abstract The fuzzy technique is used in this chapter to study the fuzzy multi-objective quadratic transportation problem (FMQTP). In addition, we look at the challenges associated with interval quadratic programming, where the cost coefficients, constraint coefficients, and interval data representations are on the right. Additionally, we construct a set of two-level models for mathematical programming used as the foundation for determining the upper and lower limits of the objective values.

Keywords Optimality · Transportation problem · Robust ranking · Fuzzy theory

4.1 Introduction

The main trend in science and engineering around the turn of the twentieth century was the distillation of complex systems in reality into precise mathematical models. Operations research (OR) was initiated in the middle of this century to apply the real-world decision-making issues. It was developed as a result of World War II military planning. After the war, the methods started to be used more often to address issues in business, industry, and society. Since then, operations research has developed into a discipline that is widely used in a variety of sectors, including finance, petrochemicals, and aviation. It has also moved to a focus on the creation of mathematical models that can be used to analyze and optimize complex systems. Operations research is the methodical use of mathematical and statistical approaches to address issues in industry, government, and education.

In Applied Mathematics, optimization is the central concept. Several disciplines are extremely interested in it. Mathematical optimization may be used to formulate

P. Anukokila (✉)
Department of Mathematics, PSG College of Arts & Science, Coimbatore, TN, India
e-mail: anuparaman@gmail.com; anukokila@psgcas.ac.in

B. Radhakrishnan
Department of Mathematics, PSG College of Technology, Coimbatore, Tamil Nadu, India
e-mail: brk.maths@psgtech.ac.in

R. Arora and S. Arora (eds.), *Optimization Techniques for Sustainable Environment Under Uncertainty*, Engineering Optimization: Methods and Applications,
https://doi.org/10.1007/978-981-96-0389-3_4

many modeling, design, control, and decision-making issues. The minimizing (or maximization) of the objectives, in light of the restrictions imposed by the issue at hand, constitutes the traditional framework for optimization. However, many design issues involve a number of objectives that must be traded off, which can result in the under-or over-achievement of numerous goals. Principle of optimization theory has been proposed by Bector et al. (2005). Optimizing techniques such as linear programming have proved useful in solving limited kinds of operating problems but have not been widely used to tackle higher level management problems. The availability of time-sharing services has permitted the use of non-optimizing techniques (e.g., simulation) to solve a wider range of problems than would be possible with analytic models.

Numerous system performance characteristics, criteria, and decision-making factors cannot always be properly quantified, nor is it always necessary. When a variable's values cannot be accurately established, it is referred to as uncertain or fuzzy. Uncertainty in values can be quantified using probability distributions. Alternatively, if they are better described by qualitative descriptors like dry or wet, hot or cold, clean or filthy, and high or low, fuzzy membership functions can be employed to quantify them. For these ambiguous or qualitative variables, quantitative optimization models can incorporate both fuzzy membership functions and probability distributions. For the purpose of solving real-world decision-making issues, fuzzy set theory is a helpful tool. It has been utilized in a number of OR strategies, including queuing theory, nonlinear programming, and linear programming. This study will focus on linear programming, the most crucial approach out of the bunch.

Zadeh, on the other hand, coined the term "fuzzy" in 1962, and formally published the classic article "Fuzzy Set Theory" in 1965 (Zadeh 1965). The oversimplified model was intended to be enhanced using fuzzy set theory in order to provide a more robust and adaptable model that could be applied to complicated human-centric systems in the real world. It is now a widely accepted fact that fuzzy structures and fuzzy theories of control have some advantages over numerous other methods whenever an intricate physical system fails to offer a set of differential or difference equations that constitute a reliable or fairly precise mathematical framework, in addition to when the system description necessitates a certain level of human comprehension in terms of language. Only in the twentieth century were the concepts of sets and functions created by mathematicians to express problems. This issue representation approach is stricter. The solutions based on this theory are frequently worthless. This issue was resolved with the aid of the fuzzy idea. Almost all mathematical, engineering, medical, and other concepts have been redefined using fuzzy sets. Therefore, it is crucial to spread awareness of these principles for our future generation.

4.2 Background and Related Work

4.2.1 Fuzzy Set

A fuzzy set is one that has "un-sharp" and ambiguous bounds. It transforms the concept of membership from a binary categorization in classical set theory to one that admits partial membership. The constraints of classical set theory are overcome by fuzzy set theory, which allows membership in a set to be a question of degree. A value between 0 and 1 represents the degree of membership in a fuzzy set; 0 indicates fully not in the set, and 1 implies completely in the set.

In one of his earliest publications on fuzzy set theory, Zadeh writes, "Ultimately, the principle of a fuzzy set serves as an ideal point concerning departure for the creation of a theoretical structure which parallels in many ways the model used in the case of conventional sets, but is more general than the former and, potentially, may prove to have an even wider range of feasibility, notably in the domains of pattern categorization and information processing." Essentially, such a framework "provides a logical approach of dealing with situations when the cause of imprecision is the absence of finely specified class membership requirements rather than the presence of random variables". The book written by Zimmermann (1978) performed an important contribution in the development of the field known as fuzzy sets decision-making and expert systems from the perspective of application in science and engineering.

Applications of this theory include artificial intelligence, computer science, health care, control engineering, decision theory, expert systems, logic, management science, operations research, pattern recognition, and robotics. Mathematical advancements have progressed to a very high level and continue to do so now. This topic focuses on optimization theory, with a particular emphasis on fuzzy linear programming. Despite the existence of many alternative techniques such as queuing theory, heuristic approaches, nonlinear programming, and so on, a review of OR literature reveals that linear programming is the most essential technique for real-world applications. The methods and applications of fuzzy mathematical programming have been established by Zimmermann (1976). Moreover he studied (Zimmermann 1987) the optimization of fuzzy systems.

4.2.2 Linear Programming

The phrase "linear optimization" can also refer to linear programming. Simply put, linear programming involves utilizing a mathematical model that is linear to optimize an output based on a set of constraints. The following symbolically expresses the generic LPP:

$$\begin{cases} \text{maximize} & \mathscr{X} = p^T y \\ \text{subject to} & \\ \mathcal{A}y \leq d & \\ y \geq 0 & \end{cases}$$

where p is the vector of profit coefficients of the objective function and d is the vector of total resources available, y is the vector of decision variables, and $\mathcal{A}$ is the matrix of technical coefficients.

In the case of linear programming, an anticipated result, such as profit maximization or cost minimization, is attained by wisely allocating scarce resources to various tasks. All of the interactions between the processes in models of linear programming are linear, satisfying both the proportionality and additive behavior criteria. Contrary to fuzzy linear programming issues, sadly, little study has been done on this significant class of problems up to this point.

One of the most reliable optimization methods used in the planning of reservoir operations is linear programming. Under particular conditions and assumptions, it can linearize nonlinear equations completely, linearize them in pieces, or solve nonlinear problems iteratively. The planning objectives often depend on how much is stored and released within the planning horizon. Most reservoir optimization problems have one of two planning objectives: either increase net management revenue or reduce operating costs while meeting all requirements (Yeh 1985). The following are the main benefits of linear programming models:

- Ability to deal with intricate problems.
- Convergence to the universal optimum is guaranteed.
- The user does not require any information to make an initial guess.
- Proficiency in carrying out sensitivity analyses.
- The utmost simplicity of problem formulation.
- Software programs that are conveniently accessible and effective, such as Lindo and Lingo.

4.2.3 Nonlinear Programming

The process of addressing an optimization issue specified by a collection of constraints, also known as equalities and inequalities, over a number of unknown real variables, together with an objective function that should be maximized or reduced, is known as nonlinear programming (NLP). Over the range of choice variables being taken into consideration, using assumptions or approximations from linear programming may also result in suitable issue representations. The objective of a universal optimization challenge is to choose n choice variables $y_1, y_2, \ldots, y_n$ in the most feasible region while optimizing the objective function representing the decision factors

$$g(y_1, y_2, \ldots, y_n).$$

If the goal function is nonlinear and/or nonlinear constraints are used to define the viable region, the issue is referred to as a nonlinear programming problem (NLPP). As a result, the generic nonlinear programming is expressed as follows in maximization form:

$$\begin{cases} \text{maximize } g(y_1, y_2, \ldots, y_n) \\ \text{subject to} \\ h_1(y_1, y_2, \ldots, y_n) \le d_1 \\ \vdots \\ h_m(y_1, y_2, \ldots, y_n) \le d_m \end{cases}$$

where each constraint function spanning h_1 through h_m is provided. The linear program that has already been addressed is an exception. In this instance, the apparent relationship is

$$\begin{cases} g(y_1, y_2, \ldots, y_n) = \sum_{j=1}^{n} p_j y_j \\ \text{and} \\ h_i(y_1, y_2, \ldots, y_n) = \sum_{j=1}^{n} e_{ij} y_j, \quad (i = 1, 2, \ldots, m). \end{cases}$$

4.2.4 Fuzzy Linear and Nonlinear Programming

According to the crisp scenario, the LPP's targeting is to maximize or minimize an objective function that is linear subject to linear constraints. The decision-maker may, however, only be able to express the objective and/or constraint functions in a "fuzzy sense," as opposed to precisely, in many real-world situations. It is preferable to utilize a fuzzy linear programming form of modeling in these circumstances. As a way to employ mathematical concepts and associated strategies to military programming and organizing problems, Dantzig suggested that "the interrelations between the activities of a large organization could be considered as a linear programming kind model and the optimizing program achieved by minimizing a linear objective function."

The design is possible to express a universal fuzzy linear programming problem (FLPP) formally in the following manner:

$$\begin{cases} \text{maximize } \widehat{\mathscr{Z}_{\mathfrak{R}}} = \tilde{p}y \\ \text{subject to} \\ \mathcal{A}y = d \\ y \ge 0 \end{cases}$$

where $d \in \mathscr{R}^m$, $y \in \mathscr{R}^n$, $\tilde{p} \in (\mathscr{F}(\mathscr{R}))^n$, $\mathcal{A} \in \mathscr{R}^{m \times n}$, $and \mathscr{R}$ is linear ranking function. Several authors (Buckly and Feuring 2000; Chanas 1983; Charnes and Cooper 1954; Delgado et al. 1989; George and Thapa 1997; Horcik 2008; Kumar et al. 2011; Mahdavi-Amiri and Nasseri 2007; Marbini and Tavana 2011; Nehi et al. 2006; Ozkok and Tiryaki 2011) have studied the fuzzy linear programming problem.

There are a lot of intriguing real-world challenges where the objective function or some of the restrictions may not be linear. Fuzzy nonlinear programming (FNLP) issues are optimization problems with nonlinearities. Abo-Sinna (2001) suggested nonlinear multi-objective decision-making under fuzziness. Xie and Jia (2012) established the nonlinear fixed charge transportation problem by minimum cost flow-based genetic algorithm. For an appropriate representation of an application as a mathematical program, nonlinearities in the form of nonlinear objective functions or nonlinear constraints are essential at other times. Following are some ways to approach a nonlinear programming issue using fuzzy coefficients:

$$\begin{cases} \text{maximize} \;\; g(\tilde{c}_1, \tilde{c}_2, \ldots, \tilde{c}_k, \; y) \\ \text{subject to} \\ h_i(\tilde{e}_1, \tilde{e}_2, \ldots, \tilde{e}_k, \; y) \leq \tilde{d}_i \\ y_i \in [l_k, u_k], \; k = 1, 2, \ldots, n, \; l_k \geq 0 \end{cases}$$

where $y = (y_1, y_2, \ldots, y_n) \in \mathscr{R}^n$ is a $n-$ dimensional real valued parameter vector, $[l_k, u_k] \subset \mathscr{R}(k = 1, 2, \ldots, n)$, $g(y)$, $h_i(y)$ are continuous arbitrary functions.

4.2.5 Transportation Problem

Transporting goods from sources to destinations is the focus of the transportation issue, a specific kind of linear programming problem. Hitchcock is credited with creating the fundamental transportation problem in Hitchcock (1941). The reason it is known as a transportation problem is because it was initially developed for figuring out the best shipping strategy. Transporting a specific product from each of m origins $i = 1, \ldots, m$ to any of n destinations $k = 1, 2, \ldots, n$ is a typical and well-known transportation problem. The initial locations are manufacturing facilities with the corresponding capacity $(c_1, c_2, \ldots, c_m)$, whereas the ending points are warehouses having the necessary levels of demand $(e_1, e_2, \ldots, e_n)$. Without sacrificing generality, one can suppose that the cost p_{ik} to transport one unit of the product under consideration from the ith source to the kth destination will be $p_{ik} \geq 0$, for all i, k.

Consequently, the standard transportation problem has the following mathematical form:

$$\begin{cases} \text{minimize } \mathscr{Z} = \sum_{i=1}^{m} \sum_{k=1}^{n} p_{ik} y_{ik} \\ \text{subject to} \\ \sum_{k=1}^{n} y_{ik} = c_i, \quad i = 1, 2, \ldots, m \\ \sum_{i=1}^{m} y_{ik} = e_k, \quad k = 1, 2, \ldots, n \\ y_{ik} \geq 0, \quad \forall\ i \text{ and } k \\ \sum_{i=1}^{m} c_i = \sum_{k=1}^{n} e_k. \end{cases}$$

The basic goal of the transportation problem is to choose a shipment schedule that minimizes the overall shipping expense while meeting supply and demand constraints. Several authors investigated the study of transportation problem (Chanas and Kuchta 1998; Dahiya and Verma 2007; Diaz 1979; Gupta and Mehlawat 2007; Jain and Saksena 2012; Poonam et al. 2012; Tada and Ishii 1996). There has been a good deal of study in the literature to find the best answer for balanced transportation challenges.

But in actual circumstances, the decision-maker must deal with an unbalanced transportation issue where the whole supply is less than the total demand. The unbalanced fuzzy transportation issue that Kumar and Kaur formulated in Kumar and Kaur (2012). The transportation issue may be solved using the simplex technique. However, an alternative approach may be developed to resolve these issues due to the unique nature of the transportation problem. It was suggested by Chanas et al. (1984), Kumar and Kaur (2011) to take a fuzzy approach to the transportation issue.

The transportation issue may be divided into linear and nonlinear categories depending on the make-up of the cost function. An illustration of a network optimization issue is transportation. With the aim of reducing overall distribution costs, it deals with the effective distribution (movement) of product (goods and services) from various supply locations (sources) with a constrained supply to various demand locations (destinations) with a defined demand. The nonlinear transportation problem is presented as follows meticulously:

- An array of n commodity suppliers with acknowledged capacity for supply and an array of m destinations with known demand.
- The nonlinear, differentiable transportation cost function spanning a unit of something from each source to each destination.

To satisfy demand at each destination while reducing overall transportation costs, we must determine the quantity of goods that must be provided from each source (and maybe sold).

Physical product distribution, often known as transportation issues, has been one of the most significant and fruitful applications of quantitative analysis in solving business difficulties. Basically, the goal is to keep the cost of moving products from one place to another as low as possible while yet meeting each arriving area's demands and allowing each shipping facility to operate to its fullest potential. Network flow issues make up a sizable portion of the practical applications of linear programming.

4.2.6 Multi-objective Transportation Problem

In actual circumstances, there are other transit issues at play. This article considers transportation issues that are characterized by several objective functions. The phrase "multi-objective transportation problem" (MOTP) designates a particular class of LPP when all of the objectives are in conflict with one another and the constraints are of the equality type. Comparable to a normal transportation issue with a MOTP, merchandise needs to be delivered from m sources to n destinations, with each having capabilities of $c_1, c_2, \ldots, c_m$ and $e_1, e_2, \ldots, e_n$.Transporting a unit of goods from i^{th} source to k^{th} destination entails a penalty of p_ik as well. Cost, delivery, or other factors could constitute the penalty. A variable y_{ik} represents the unknown quantity to be shipped from i^{th} source to k^{th} destination. A mathematical model of MOTP with w objectives, m sources, and n destinations can be written as

$$\begin{cases} \text{minimize } \mathscr{Z}_w = \sum_{i=1}^{m} \sum_{k=1}^{n} p_{ik}^{w} y_{ik}, \quad w = 1, 2, \ldots, q \\ \text{subject to} \\ \sum_{k=1}^{n} y_{ik} \leq c_i, \quad i = 1, 2, \ldots, m \\ \sum_{i=1}^{m} y_{ik} \leq e_k, \quad k = 1, 2, \ldots, n \\ y_{ik} \geq 0, \quad \text{for all } i \text{ and } k. \end{cases}$$

The subscript on $\mathscr{Z}_w$ and superscript on p_{ik}^{w} are related to the wth penalty criterion. It may be assumed that $c_i \geq 0$ and $e_k \geq 0$, for all i, k and the equilibrium condition $\sum_{i=1}^{m} c_i = \sum_{k=1}^{n} e_k$ is satisfied. Several papers (Bit et al. 1993; Pramanik and Roy 2008; Preetvanti and Saxena 2003; Tada and Ishii 1996) appeared on various types of multi-objective transportation problem with fuzzy parameters in fuzzy approach. For figuring out MOTP, there are numerous distinct ways, including geometric programming, interactive fuzzy multi-objective linear programming, fuzzy goal programming, fuzzy goal programming, fuzzy programming methodology, and fuzzy goal programming. Nevertheless, the notion of fuzzy goal programming strategy has been emphasized in this study.

4.2.7 Solid Transportation Problem

Haley originally introduced the concept of the solid transportation problem (STP) in Haley (1962) and explored the multi-index problem in Haley (1963), which takes into account three different types of constraints: source constraint, destination constraint, and conveyance constraint. As there is only one vehicle, the STP degenerates toward the conventional transportation problem. It is regarded as a unique instance of LPP. The fact that numerous industrial problems take on this particular shape makes it necessary to take this particular type of transportation problem into consideration. It should be emphasized that when a variety of conveyances are available for the shipment of products, the STP becomes necessary. By focusing on a single mode of transportation, the STP can be transformed into a traditional transportation problem. It has been extensively studied by several authors (Ida et al. 1996; Li et al. 1997; Patel and Tripathy 1989; Yang and Liu 2007).

Take the following aspects of STP into consideration: The decision-maker of a transportation firm will calculate the transportation quantity based on preliminary research in order to break into an unfamiliar market. The decision-maker frequently calculates the shipping capacity as a sufficiently high figure to guarantee the supply out of concern for the company's reputation. A universe, symbolized by the symbol d_k, can be thought of as an adequately huge integer. Occasionally, the decision-maker is not involved in the detailed analysis or actual implementation of the transit plan.

4.2.8 Quadratic Programming

Quadratic programming constitutes one of the most significant techniques in operations research. An optimization circumstance with a quadratic objective function and a linear restriction is known as a quadratic programming problem. Quadratic programming is sometimes recognized as a separate subject due to its wide range of applications. The cost per unit of a product sent from a given source to a given destination is constant in the linear transportation issue (also known as the "ordinary transportation problem"), regardless of the quantity shipped. The mileage (distance) between every source and every destination is typically assumed to be constant.

The subsequent one is an assortment of quadratic programming with fuzzy numbers (QPFN) problems:

$$\begin{cases} \text{minimize } \frac{1}{2}y^T\widetilde{\mathbb{P}}y + y^T\widetilde{\mathbb{D}} \\ \text{subject to} \\ \tilde{A}y \leq \tilde{d}. \end{cases}$$

A mathematical modeling strategy called quadratic programming aims to maximize the use of finite resources. Quadratic programming models are typically created in real-world applications to determine a plan of action for the future. Making a forecast of the future circumstances would be employed to calculate the parameter values, which entails a certain amount of risk. It has produced a variety of intriguing uses as well as various beneficial outcomes. Typically, a second-order quadratic system is used to approximate optimization problems with nonlinear objective functions, and then a normal quadratic programming approach is used to solve them sequentially. Several authors (Arora and Khurana 2004; Wu and Liu 2012; Wu and Zhang 2008) studied goal programming problems.

One of the effective methods for boosting the effectiveness and productivity of commercial enterprises and governmental agencies is quadratic programming. When real situations are considered, vagueness appears in a natural way, and hence it makes perfect sense to think of fuzzy quadratic programming problems. Igeartaigh (1982) given fuzzy transportation algorithm, and Lau et al. (2009) proposed multi-objective evolutionary algorithm for transportation problem. The context of their discussion by Qiao et al. (1994) point out different linear programming models incorporating fuzzy random variables together with fuzzy random coefficients.

Quadratic programming can be viewed both as a special case of the nonlinear programming and a generalization of the linear programming. Pal et al. (2011) presented a fuzzy goal programming procedure for solving quadratic bi-level programming. Duality in fuzzy quadratic programming was derived by Gupta and Dangar (2010). Yager developed the scale setting and broader concept for interval valued fuzzy sets and their application in uncertainty measures in his work reported in Yager (2008).

The fuzzy quadratic programming task is converted into two two-level mathematical programs built around Zadeh's extension principle, Zadeh (1978), to determine the upper and lower bounds of the objective value at possibility level β. Enumerating various β values leads to the numerical derivation of the membership function of the fuzzy objective value. Further, Liu (2009) proposes an effective solution method to solving a class of fuzzy quadratic programming problems. Punnen and Zhang (2011) provides quadratic Bottleneck problems. Numerical solution method for quadratic programming was discussed by Li and Tian (2008), Liu and Wang (2007). Bellman and Zadeh (1970) put out the maximization decision notion. By incorporating this idea of fuzzy sets, Zimmermann (1992) became the first to do so in mathematical programs.

4.3 Preliminaries

We need some basic definitions and properties of fuzzy random variables related to this chapter. Let $\mathscr{F}_0(\mathcal{R})$ stand for the collection of all compact fuzzy numbers over $\mathscr{R}$.

Definition 4.1 For $a_\beta = [a_\beta^{\mathcal{L}}, a_\beta^{\mathcal{U}}]$ and $b_\beta = [b_\beta^{\mathcal{L}}, b_\beta^{\mathcal{U}}]$, we define

(1) $a_\beta + b_\beta = [a_\beta^{\mathcal{L}}, a_\beta^{\mathcal{U}}] + [b_\beta^{\mathcal{L}}, b_\beta^{\mathcal{U}}] = [a_\beta^{\mathcal{L}} + b_\beta^{\mathcal{L}},\ a_\beta^{\mathcal{U}} + b_\beta^{\mathcal{U}}]$.
(2) $a_\beta - b_\beta = [a_\beta^{\mathcal{L}}, a_\beta^{\mathcal{U}}] - [b_\beta^{\mathcal{L}}, b_\beta^{\mathcal{U}}] = [a_\beta^{\mathcal{L}} - b_\beta^{\mathcal{L}},\ a_\beta^{\mathcal{U}} - b_\beta^{\mathcal{U}}]$.
(3) $a_\beta \cdot b_\beta = [a_\beta^{\mathcal{L}}, a_\beta^{\mathcal{U}}] \cdot [b_\beta^{\mathcal{L}}, b_\beta^{\mathcal{U}}] = [a_\beta^{\mathcal{L}} b_\beta^{\mathcal{L}} \wedge a_\beta^{\mathcal{L}} b_\beta^{\mathcal{U}} \wedge\ a_\beta^{\mathcal{U}} b_\beta^{\mathcal{L}} \wedge a_\beta^{\mathcal{U}} b_\beta^{\mathcal{U}},\ a_\beta^{\mathcal{L}} b_\beta^{\mathcal{L}} \vee a_\beta^{\mathcal{L}}$ $b_\beta^{\mathcal{U}} \vee a_\beta^{\mathcal{U}} b_\beta^{\mathcal{L}} \vee a_\beta^{\mathcal{U}} b_\beta^{\mathcal{U}}]$.
(4) The order relation "$\leq$" is defined by

$$[a_\beta^{\mathcal{L}}, a_\beta^{\mathcal{U}}] \leq [b_\beta^{\mathcal{L}}, b_\beta^{\mathcal{U}}] \text{ if and only if } a_\beta^{\mathcal{L}} \leq b_\beta^{\mathcal{L}},\ \ a_\beta^{\mathcal{U}} \leq b_\beta^{\mathcal{U}}.$$

(5) Let $[a_{\beta i}^{\mathcal{L}},\ a_{\beta i}^{\mathcal{U}}] \subset \mathscr{R},\ i \in \mathscr{I},\ \mathscr{I}$ be the index set, then

$$\wedge_{i \in \mathscr{I}} [a_{\beta i}^{\mathcal{L}}, a_{\beta i}^{\mathcal{U}}] = [\wedge_{i \in \mathscr{I}} a_{\beta i}^{\mathcal{L}}, \wedge_{i \in \mathscr{I}} a_{\beta i}^{\mathcal{U}}] \ \ if \ \ \wedge_{i \in \mathscr{I}} a_{\beta i}^{\mathcal{L}} > -\infty.$$
$$\vee_{i \in \mathscr{I}} [a_{\beta i}^{\mathcal{L}}, a_{\beta i}^{\mathcal{U}}] = [\vee_{i \in \mathscr{I}} a_{\beta i}^{\mathcal{L}}, \vee_{i \in \mathscr{I}} a_{\beta i}^{\mathcal{U}}] \ \ if \ \ \vee_{i \in \mathscr{I}} a_{\beta i}^{\mathcal{L}} > \infty.$$

Definition 4.2 Let $\{\tilde{a}_i | i \in \mathscr{I}\} \subset \mathscr{F}_0(\mathscr{R}),\ \beta \in (0, 1]$, then

(1) $\tilde{g} = \wedge_{i \in \mathscr{I}}\ \tilde{a}_i$ is defined by a fuzzy number $\tilde{a} \in \mathscr{F}_0(\mathscr{R})$ such that $g_\beta = \wedge_{i \in \mathscr{I}} (a_i)_\beta$.
(2) $\tilde{h} = \wedge_{i \in \mathscr{I}}\ \tilde{a}_i$ is defined by a fuzzy number $\tilde{a} \in \mathscr{F}_0(\mathscr{R})$ such that $h_\beta = \wedge_{i \in \mathscr{I}} (a_i)_\beta$.

Definition 4.3 Let $(\Omega, \mathcal{A}, \mathscr{P})$ be a probability measure space. A mapping $\tilde{a} : \Omega \to \mathscr{F}_0(\mathscr{R})$ is called a fuzzy random variable on $(\Omega, \mathcal{A})$ if for any $\beta \in (0, 1]$

$$a_\beta(\eta) = \left\{ y | y \in \mathscr{R}, \tilde{a}(\eta)(y) \geq \beta \right\} = [a_\beta^{\mathcal{L}}(\eta), a_\beta^{\mathcal{U}}(\eta)]$$

is a random interval, that is $a_\beta^{\mathcal{L}}(\eta),\ a_\beta^{\mathcal{U}}(\eta)$ are two random variables on $(\Omega, \mathcal{A})$. Denote the set of all fuzzy random variables on $(\Omega, \mathcal{A})$ by $\mathscr{F}\mathscr{R}(\Omega)$ (where $\eta \in \Omega$). The algebraic operation $*$ on $\mathscr{F}\mathscr{R}(\Omega)$ may be defined by

$$(\tilde{u} * \tilde{v})(\eta) = \tilde{u}(\eta) * \tilde{v}(\eta), \text{ for any } \eta \in \Omega.$$

Also, for any $\tilde{u},\ \tilde{v} \in \mathscr{F}\mathscr{R}(\Omega)$

$$(\tilde{u} \leq \tilde{v})(\eta) = \tilde{u}(\eta) \leq \tilde{v}(\eta), \text{ for any } \eta \in \Omega.$$

4.4 Problem Formulation

4.4.1 Interval Quadratic Transportation Problem

This section describes the linear programming model for MOTP in interval numbers including crisp functions. Transporting a unit of the product from source i toward destination j through the conveyance k is accompanied by a fine or cost of $v_{ijk} \geq 0$.

We use the fuzzy random multi-objective quadratic transportation issue described below as an instance

$$\begin{cases} \text{maximize } \widehat{\mathscr{Z}^v}(\tilde{y}) = \sum_{i=1}^{m}\sum_{j=1}^{n} p_{ij}y_{ij} + \frac{1}{2}\sum_{i=1}^{m}\sum_{j=1}^{n}\sum_{k=1}^{l} v_{ijk}y_{ijk} \\ \text{subject to} \\ \sum_{j=1}^{n}\sum_{k=1}^{l} y_{ijk} = \mathscr{A}_i, \quad \text{for } i = 1, 2, \ldots, m \\ \sum_{i=1}^{m}\sum_{k=1}^{l} y_{ijk} = \mathscr{B}_j, \quad \text{for } j = 1, 2, \ldots, n \\ \sum_{i=1}^{m}\sum_{j=1}^{n} y_{ijk} = \mathscr{E}_k, \quad \text{for } k = 1, 2, \ldots, l \\ y_{ijk} \geq 0, \text{ for all } i, j, k \end{cases} \tag{4.1}$$

where $p_{ij} \in [p_{ij}^{\mathcal{U}}, p_{ij}^{\mathcal{L}}]$, $\mathscr{A}_i \in [\mathscr{A}_i^{\mathcal{L}}, \mathscr{A}_i^{\mathcal{U}}]$, $\mathscr{B}_j \in [\mathscr{B}_j^{\mathcal{L}}, \mathscr{B}_j^{\mathcal{U}}]$, and $\mathscr{E}_k = [\mathscr{E}_k^{\mathcal{L}}, \mathscr{E}_k^{\mathcal{U}}]$ are the interval inter parts of p_{ij}, $\mathscr{A}_i$, $\mathscr{B}_j$, $\mathscr{E}_k$ respectively. For equality constraint of the "=" form. Locating the upper and lower bounds of equation (4.1)'s objective values is sufficient to determine the range containing the objective values. The values p_{ij}, $\mathscr{A}_i$, $\mathscr{B}_j$, $\mathscr{E}_k$ attain the largest value for $\mathscr{Z}$ that can be determined from the following two-level mathematical programming model:

$$\begin{cases} \mathscr{Z}_v^{\mathcal{U}} = \text{maximize}_{(p,\mathscr{A},\mathscr{B},\mathscr{E})\in s} \text{ minimize}_y \mathscr{Z} \\ = \sum_{i=1}^{m}\sum_{j=1}^{n} p_{ij}y_{ij} + \frac{1}{2}\sum_{i=1}^{m}\sum_{j=1}^{n}\sum_{k=1}^{l} v_{ijk}y_{ijk} \\ \text{subject to} \\ \sum_{j=1}^{n}\sum_{k=1}^{l} y_{ijk} = \mathscr{A}_i, \quad \text{for } i = 1, 2, \ldots, m \\ \sum_{i=}^{m}\sum_{k=1}^{l} y_{ijk} = \mathscr{B}_j, \quad \text{for } j = 1, 2, \ldots, n \\ \sum_{i=1}^{m}\sum_{j=1}^{n} y_{ijk} = \mathscr{E}_k, \quad \text{for } k = 1, 2, \ldots, l \\ y_{ijk} \geq 0, \text{ for all } i, j, k. \end{cases} \tag{4.2}$$

A two-level mathematical program is formulated by replacing the outer program of equation (4.2) from minimize to maximize

$$\begin{cases} Z_v^{\mathcal{L}} = \text{minimize}_{(p,\mathscr{A},\mathscr{B},\mathscr{E})\in s} \text{ minimize}_y \mathscr{Z} \\ = \sum_{i=1}^{m}\sum_{j=1}^{n} p_{ij} y_{ij} + \frac{1}{2}\sum_{i=1}^{m}\sum_{j=1}^{n}\sum_{k=1}^{l} v_{ijk} y_{ijk} \\ \text{subject to} \\ \sum_{j=1}^{n}\sum_{k=1}^{l} y_{ijk} = \mathscr{A}_i, \quad \text{for } i = 1, 2, \ldots, m \\ \sum_{i=1}^{m}\sum_{k=1}^{l} y_{ijk} = \mathscr{B}_j, \quad \text{for } j = 1, 2, \ldots, n \\ \sum_{i=1}^{m}\sum_{j=1}^{n} y_{ijk} = \mathscr{E}_k, \quad \text{for } k = 1, 2, \ldots, l \\ y_{ijk} \geq 0, \text{ for all } i, j, k. \end{cases} \tag{4.3}$$

4.4.2 *Weighting Problem of FMQTP*

Definition 4.4 A point y^* is called fuzzy efficient solution of FMQTP problem if $\mathscr{Z}^v(y^*) \leq \mathscr{Z}^v(y)$ with $\mathscr{Z}^v(y^*) < \mathscr{Z}^v(y)$ holds for at least one $v = 1, 2, \ldots, \mathscr{Q}$.

Model 1: Consider the multi-objective weighting quadratic transportation problem:

$$\begin{cases} \text{minimize}_y \left(\sum_{r=1}^{k} w^r \left(\sum_{i=1}^{m}\sum_{j=1}^{n} p_{ij}^r y_{ij} + \frac{1}{2}\sum_{i=1}^{m}\sum_{j=1}^{n}\sum_{k=1}^{l} v_{ijk}^r y_{ijk}\right)\right) \\ = \left(w^1 \sum_{i=1}^{m}\sum_{j=1}^{n} p_{ij}^1 y_{ij} + \frac{1}{2}\sum_{i=1}^{m}\sum_{j=1}^{n}\sum_{k=1}^{l} v_{ijk}^1 y_{ijk}\right) \\ + \ldots + \left(w^r \sum_{i=1}^{m}\sum_{j=1}^{n} p_{ij}^k y_{ij} + \frac{1}{2}\sum_{i=1}^{m}\sum_{j=1}^{n}\sum_{k=1}^{l} q_{ijk}^k y_{ijk}\right) \\ \text{subject to} \\ \sum_{j=1}^{n}\sum_{k=1}^{l} y_{ijk} = \mathscr{A}_i \\ \sum_{i=1}^{m}\sum_{k=1}^{l} y_{ijk} = \mathscr{B}_j \\ \sum_{i=1}^{m}\sum_{j=1}^{n} y_{ijk} = \mathscr{E}_k \text{ and } \mathscr{M} = y \in \mathscr{R}^n, \quad y \geq 0 \end{cases} \tag{4.4}$$

where $w = \left\{w^r \in \Omega : w^r \geq 0, \ \sum_{r=1}^{k} w^r = 1\right\}$ is the contexts to ensure that the weights are normalized.

Definition 4.5 A point $\mathscr{X}^*(w^*)$ is called fuzzy optimal solution of the problem. Model 1. If for each $y \in \mathscr{M}$

$$\sum_{r=1}^{k} w^r \left(\sum_{i=1}^{m} \sum_{j=1}^{n} p_{ij} y_{ij} + \frac{1}{2} \sum_{i=1}^{m} \sum_{j=1}^{n} \sum_{k=1}^{l} v_{ijk} y^*_{ijk} \right)$$
$$\leq \sum_{r=1}^{k} w^r \left(\sum_{i=1}^{m} \sum_{j=1}^{n} p_{ij} y_{ij} + \frac{1}{2} \sum_{i=1}^{m} \sum_{j=1}^{n} \sum_{k=1}^{l} v_{ijk} y_{ijk} \right).$$

Theorem 4.1 *Let* $a,\ b \in \mathscr{F}_0(\mathbb{R})$, *then, for any* $\beta \in [0, 1]$ *the* β*−level cuts of* $v_{ijk},\ \mathscr{A}_i,\ \mathscr{B}_j,\ \mathscr{E}_k,\ p_{ijk},\ y_{ijk} \in \mathscr{F}\mathbb{R}(\Omega)$ *in random interval are denoted by*

$$\begin{aligned}
(v^r_{ijk})_\beta &= [(v^r_{ijk})^{\mathcal{L}}_\beta, (v^r_{ijk})^{\mathcal{U}}_\beta] \\
(\mathscr{A}_i)_\beta &= [(\mathscr{A}_i)^{\mathcal{L}}_\beta, (\mathscr{A}_i)^{\mathcal{U}}_\beta] \\
(\mathscr{B}_j)_\beta &= [(\mathscr{B}_j)^{\mathcal{L}}_\beta, (\mathscr{B}_j)^{\mathcal{U}}_\beta] \\
(\mathscr{E}_{ij})_\beta &= [(\mathscr{E}_{ij})^{\mathcal{L}}_\beta, (\mathscr{E}_{ij})^{\mathcal{U}}_\beta] \\
(y_{ijk})_\beta &= [(y_{ijk})^{\mathcal{L}}_\beta, (y_{ijk})^{\mathcal{U}}_\beta].
\end{aligned}$$

Theorem 4.2 *Suppose that* $\mathscr{A} \geq 0,\ \mathscr{B} \geq 0,\ \mathscr{E} \geq 0,\ v^r \geq 0,$ *and* $\mathscr{C}^r \geq 0,\ r = 1, 2, \ldots, k$. *If* $\mathscr{X}^*$ *is a fuzzy pseudo random optimal solution of model 1, then for any* $\beta \in (0, 1]$, *we have*

$$\begin{cases}
\mathscr{Z}^{\mathcal{L}}_\beta = \text{minimize} \sum_{r=1}^{k} w^r \left(\sum_{i=1}^{m} \sum_{j=1}^{n} (p_{ij})^{\mathcal{L}}_\beta y_{ij} + \frac{1}{2} \sum_{i=1}^{m} \sum_{j=1}^{n} \sum_{k=1}^{l} (v_{ijk})^{\mathcal{L}}_\beta y_{ijk} \right) \\
\mathscr{Z}^{\mathcal{U}}_\beta = \text{minimize} \sum_{r=1}^{k} w^r \left(\sum_{i=1}^{m} \sum_{j=1}^{n} (p_{ij})^{\mathcal{U}}_\beta y_{ij} + \frac{1}{2} \sum_{i=1}^{m} \sum_{j=1}^{n} \sum_{k=1}^{l} (v_{ijk})^{\mathcal{U}}_\beta y_{ijk} \right)
\end{cases}$$

where $Z = \text{minimize} \sum_{r=1}^{k} w^r \left(\sum_{i=1}^{m} \sum_{j=1}^{n} (p_{ij}) y_{ij} + \frac{1}{2} \sum_{i=1}^{m} \sum_{j=1}^{n} \sum_{k=1}^{l} v_{ijk} y_{ijk} \right).$

Proof Suppose that y^* is a solution of Model 1, then $y^* \in \mathscr{M}$ and

$$\sum_{r=1}^{k} w^r \left(\sum_{i=1}^{m} \sum_{j=1}^{n} (p_{ij}) y^*_{ij} + \frac{1}{2} \sum_{i=1}^{m} \sum_{j=1}^{n} \sum_{k=1}^{l} v_{ijk} y^*_{ijk} \right)$$
$$\leq \sum_{r=1}^{k} w^r \left(\sum_{i=1}^{m} \sum_{j=1}^{n} (p_{ij}) y_{ij} + \frac{1}{2} \sum_{i=1}^{m} \sum_{j=1}^{n} \sum_{k=1}^{l} v_{ijk} y_{ijk} \right). \tag{4.5}$$

L.H.S. of (4.5) becomes

$$\sum_{r=1}^{k} w^r \left(\sum_{i=1}^{m} \sum_{j=1}^{n} (p_{ij})^r y_{ij}^* + \frac{1}{2} \sum_{i=1}^{m} \sum_{j=1}^{n} \sum_{k=1}^{l} v_{ijk} y_{ijk}^* \right)$$

$$
\begin{aligned}
&= \sum_{r=1}^{k} w^r \left(\sum_{i=1}^{m} \sum_{j=1}^{n} [(p_{ij}^r)_\beta^{\mathcal{L}}, (p_{ij}^r)_\beta^{\mathcal{U}}][(y_{ij}^*)_\beta^{\mathcal{L}}, (y_{ij}^*)_\beta^{\mathcal{U}}] \right) \\
&+ \sum_{r=1}^{k} w^r \left(\frac{1}{2} \sum_{i=1}^{m} \sum_{j=1}^{n} \sum_{k=1}^{l} [(v_{ijk}^r)_\beta^{\mathcal{L}}, (v_{ijk}^r)_\beta^{\mathcal{U}}][(y_{ijk}^*)_\beta^{\mathcal{L}}, (y_{ijk}^*)_\beta^{\mathcal{U}}] \right) \\
&= \sum_{r=1}^{k} w^r \left(\sum_{i=1}^{m} \sum_{j=1}^{n} (p_{ij}^r)_\beta^{\mathcal{L}}, (y_{ij}^*)_\beta^{\mathcal{L}} + \frac{1}{2} \sum_{i=1}^{m} \sum_{j=1}^{n} \sum_{k=1}^{l} (v_{ijk}^r)_\beta^{\mathcal{L}} (y_{ijk}^*)_\beta^{\mathcal{L}} \right) \\
&+ \left(\sum_{i=1}^{m} \sum_{j=1}^{n} (p_{ij}^r)_\beta^{\mathcal{U}}, (y_{ij}^*)_\beta^{\mathcal{U}} + \frac{1}{2} \sum_{i=1}^{m} \sum_{j=1}^{n} \sum_{k=1}^{l} (v_{ijk}^r)_\beta^{\mathcal{U}} (y_{ijk}^*)_\beta^{\mathcal{U}} \right) \\
&= \sum_{r=1}^{k} w^r \left(\frac{1}{2} \sum_{i=1}^{m} \sum_{j=1}^{n} \sum_{k=1}^{l} [(v_{ijk}^r)_\beta^{\mathcal{L}}, (v_{ijk}^r)_\beta^{\mathcal{U}}][(y_{ijk}^*)_\beta^{\mathcal{L}}, (y_{ijk}^*)_\beta^{\mathcal{U}}] \right) \\
&+ \sum_{r=1}^{k} w^r \left(\sum_{i=1}^{m} \sum_{j=1}^{n} [(p_{ij}^r)_\beta^{\mathcal{L}}, (p_{ij}^r)_\beta^{\mathcal{U}}][(y_{ij}^*)_\beta^{\mathcal{L}}, (y_{ij}^*)_\beta^{\mathcal{U}}] \right) \\
&= \sum_{r=1}^{k} w^r \left(\frac{1}{2} \sum_{i=1}^{m} \sum_{j=1}^{n} \sum_{k=1}^{l} [(v_{ijk}^{r\mathcal{L}})_\beta, (v_{ijk}^{r\mathcal{U}})_\beta][(y_{ijk}^*)_\beta^{\mathcal{L}}, (y_{ijk}^*)_\beta^{\mathcal{U}}] \right) \\
&+ \sum_{r=1}^{k} w^r \left(\sum_{i=1}^{m} \sum_{j=1}^{n} [(p_{ij}^{r\mathcal{L}})_\beta, (p_{ij}^{r\mathcal{U}})_\beta][(y_{ij}^*)_\beta^{\mathcal{L}}, (y_{ij}^*)_\beta^{\mathcal{U}}] \right) \\
&= \sum_{r=1}^{k} w^r \left(\sum_{i=1}^{m} \sum_{j=1}^{n} p_{ij}^r y_{ij}^* + \sum_{i=1}^{m} \sum_{j=1}^{n} \sum_{k=1}^{l} v_{ijk} y_{ijk}^* \right).
\end{aligned}
\tag{4.6}
$$

R.H.S. of Eq. (4.5) becomes

$$\sum_{r=1}^{k} w^r \left(\sum_{i=1}^{m} \sum_{j=1}^{n} (p_{ij}) y_{ij} + \frac{1}{2} \sum_{i=1}^{m} \sum_{j=1}^{n} \sum_{k=1}^{l} v_{ijk} y_{ijk} \right)$$

$$= \sum_{r=1}^{k} w^r \left(\sum_{i=1}^{m} \sum_{j=1}^{n} [(p_{ij}^r)_\beta^{\mathcal{L}}, (p_{ij}^r)_\beta^{\mathcal{U}}][(y_{ij})_\beta^{\mathcal{L}}, (y_{ij})_\beta^{\mathcal{U}}] \right)$$

$$+ \sum_{r=1}^{k} w^r \left(\frac{1}{2} \sum_{i=1}^{m} \sum_{j=1}^{n} \sum_{k=1}^{l} [(v_{ijk}^r)_\beta^{\mathcal{L}}, (v_{ijk}^r)_\beta^{\mathcal{U}}][(y_{ijk})_\beta^{\mathcal{L}}, (y_{ijk})_\beta^{\mathcal{U}}] \right)$$

$$= \sum_{r=1}^{k} w^r \left(\sum_{i=1}^{m} \sum_{j=1}^{n} (p_{ij}^r)_\beta^{\mathcal{L}}, (y_{ij})_\beta^{\mathcal{L}} + \frac{1}{2} \sum_{i=1}^{m} \sum_{j=1}^{n} \sum_{k=1}^{l} (v_{ijk}^r)_\beta^{\mathcal{L}} (y_{ijk})_\beta^{\mathcal{L}} \right)$$

$$+ \left(\sum_{i=1}^{m} \sum_{j=1}^{n} (p_{ij}^r)_\beta^{\mathcal{U}}, (y_{ij})_\beta^{\mathcal{U}} + \frac{1}{2} \sum_{i=1}^{m} \sum_{j=1}^{n} \sum_{k=1}^{l} (v_{ijk}^r)_\beta^{\mathcal{U}} (y_{ijk})_\beta^{\mathcal{U}} \right)$$

$$= \sum_{r=1}^{k} w^r \left(\frac{1}{2} \sum_{i=1}^{m} \sum_{j=1}^{n} \sum_{k=1}^{l} [(v_{ijk}^r)_\beta^{\mathcal{L}}, (v_{ijk}^r)_\beta^{\mathcal{U}}][(y_{ijk})_\beta^{\mathcal{L}}, (y_{ijk})_\beta^{\mathcal{U}}] \right)$$

$$+ \sum_{r=1}^{k} w^r \left(\sum_{i=1}^{m} \sum_{j=1}^{n} [(p_{ij}^r)_\beta^{\mathcal{L}}, (p_{ij}^r)_\beta^{\mathcal{U}}][(y_{ij})_\beta^{\mathcal{L}}, (y_{ij})_\beta^{\mathcal{U}}] \right)$$

$$= \sum_{r=1}^{k} w^r \left(\frac{1}{2} \sum_{i=1}^{m} \sum_{j=1}^{n} \sum_{k=1}^{l} [(v_{ijk}^{r\mathcal{L}})_\beta, (v_{ijk}^{r\mathcal{U}})_\beta][(y_{ijk})_\beta^{\mathcal{L}}, (y_{ijk})_\beta^{\mathcal{U}}] \right)$$

$$+ \sum_{r=1}^{k} w^r \left(\sum_{i=1}^{m} \sum_{j=1}^{n} [(p_{ij}^{r\mathcal{L}})_\beta, (p_{ij}^{r\mathcal{U}})_\beta][(y_{ij})_\beta^{\mathcal{L}}, (y_{ij})_\beta^{\mathcal{U}}] \right)$$

$$= \sum_{r=1}^{k} w^r \left(\sum_{i=1}^{m} \sum_{j=1}^{n} p_{ij}^r y_{ij} + \sum_{i=1}^{m} \sum_{j=1}^{n} \sum_{k=1}^{l} v_{ijk} y_{ijk} \right) \tag{4.7}$$

and from Eqs. (4.6) and (4.7) we obtain

$$\begin{cases} \sum_{r=1}^{k} w^r \left(\sum_{i=1}^{m} \sum_{j=1}^{n} p_{ij}^{\mathcal{L}} y_{ij}^{\mathcal{L}} + \frac{1}{2} \sum_{i=1}^{m} \sum_{j=1}^{n} \sum_{k=1}^{l} v_{ijk}^{\mathcal{L}} y_{ijk}^{\mathcal{L}} \right) \\ and \\ \sum_{r=1}^{k} w^r \left(\sum_{i=1}^{m} \sum_{j=1}^{n} p_{ij}^{\mathcal{U}} y_{ij}^{\mathcal{U}} + \frac{1}{2} \sum_{i=1}^{m} \sum_{j=1}^{n} \sum_{k=1}^{l} v_{ijk}^{\mathcal{U}} y_{ijk}^{\mathcal{U}} \right) \end{cases} \tag{4.8}$$

and Eq. (4.8) shows that the part of the theorem is correct too. This completes the proof. □

The two-level program of equation (4.2) can be used to compute the upper bound of the objective value of the interval quadratic program of equation (4.1). However, because the outer program and inner program have opposite orientations for optimal, i.e., one for maximization and another for minimizing, the solution to equation (4.2) is not so simple. The Lagrangian dual problem of (4.1) is to maximize $\Lambda(\eta, \zeta)$ over $\eta,\ \zeta \geq 0$, where

$$\begin{aligned} \Lambda(\eta, \zeta) = & \sum_{i=1}^{m} \sum_{j=1}^{n} p_{ij} y_{ij} + \frac{1}{2} \sum_{i=1}^{m} \sum_{j=1}^{n} \sum_{k=1}^{l} v_{ijk} y_{ijk} \\ & + \sum_{i=1}^{m} \eta_i \left(\sum_{j=1}^{n} a_{ij} y_j - b_i \right) - \sum_{j=1}^{n} \zeta_j y_j . \end{aligned} \tag{4.9}$$

Since (4.1) is symmetric and positive semi-definite, for a given λ and δ, the function

$$\sum_{i=1}^{m} \sum_{j=1}^{n} p_{ij} y_{ij} + \frac{1}{2} \sum_{i=1}^{m} \sum_{j=1}^{n} \sum_{k=1}^{l} v_{ijk} y_{ijk} + \sum_{i=1}^{m} \eta_i \left(\sum_{j=1}^{n} a_{ij} y_j - b_i \right) - \sum_{j=1}^{n} \zeta_j y_j$$

is convex.

A necessary and sufficient condition for a minimum is that the gradient must vanish, that is

$$\sum_{i=1}^{m} p_{ij} y_i + \sum_{i=1}^{m} \sum_{k=1}^{l} v_{ijk} y_{ik} + \sum_{i=1}^{m} \eta_i (a_{ij}) - \zeta_j = 0, \quad j = 1, 2, \ldots, n. \tag{4.10}$$

Thus, the dual form of the inner program in (4.2) can be written as follows:

$$
\begin{cases}
\text{maximize}_{(y,\eta,\zeta)}\mathscr{L} = \sum_{i=1}^{m}\sum_{j=1}^{n} p_{ij}y_{ij} + \frac{1}{2}\sum_{i=1}^{m}\sum_{j=1}^{n}\sum_{k=1}^{l} v_{ijk}y_{ijk} \\
+\sum_{i=1}^{m}\eta_i\left(\sum_{j=1}^{n} a_{ij}y_j - b_i\right) - \sum_{j=1}^{n}\zeta_j y_j \\
\text{subject to} \\
\sum_{i=1}^{m} p_{ij}y_i + \sum_{i=1}^{m}\sum_{k=1}^{l} v_{ijk}y_{ik} + \sum_{i=1}^{m}\lambda_i(a_{ij}) - \zeta_j = 0 \\
\sum_{j=1}^{n}\sum_{k=1}^{l} y_{ijk} = \mathscr{A}_i, \quad \text{for } i = 1, 2, \ldots, m \\
\sum_{i=1}^{m}\sum_{k=1}^{l} y_{ijk} = \mathscr{B}_j, \quad \text{for } j = 1, 2, \ldots, n \\
\sum_{i=1}^{m}\sum_{j=1}^{n} y_{ijk} = \mathscr{E}_k, \quad \text{for } k = 1, 2, \ldots, l \\
y_{ijk} \geq 0, \text{ for all } i,\ j,\ k \quad (\eta_i, \zeta_j) \geq 0.
\end{cases}
\tag{4.11}
$$

Now from (4.10), we have

$$
\sum_{j=1}^{n}\sum_{i=1}^{m} p_{ij}y_i + \sum_{i=1}^{m}\eta_i\left(\sum_{j=1}^{n} a_{ij}\right)y_j - \sum_{j=1}^{n}\zeta_j y_j = -\sum_{j=1}^{n}\sum_{i=1}^{m}\sum_{k=1}^{l} v_{ijk}y_{ijk}.
$$

Substituting this in (4.11), we have

$$
\begin{cases}
\text{maximize}_{(y,\eta,\zeta)}\mathscr{L} = -\frac{1}{2}\sum_{i=1}^{m}\sum_{j=1}^{n}\sum_{k=1}^{l} v_{ijk}y_{ijk} - \sum_{i=1}^{m} b_i\eta_i \\
\text{subject to} \\
\sum_{i=1}^{m}\sum_{k=1}^{l} v_{ijk}y_{ik} + \sum_{i=1}^{m}\eta_i a_{ij} - \zeta_j = -\sum_{i=1}^{m} p_{ij}y_i \\
\sum_{j=1}^{n}\sum_{k=1}^{l} y_{ijk} = \mathscr{A}_i \\
\sum_{i=1}^{m}\sum_{k=1}^{l} y_{ijk} = \mathscr{B}_j \\
\sum_{i=1}^{m}\sum_{j=1}^{n} y_{ijk} = \mathscr{E}_k \\
y_{ijk} \geq 0, \quad \text{for all } i,\ j,\ k \\
(\eta_i, \zeta_j) \geq 0, \quad i = 1, 2, \ldots, m, \quad j = 1, 2, \ldots, n.
\end{cases}
\tag{4.12}
$$

4.5 The Solution Procedure

The writer has designed the fuzzy solid quadratic transportation issue solution approach in this section. Based on the extension principle, the membership function $\mu_{\widehat{z}}$ can be defined as

$$\begin{aligned}\mu_{\widehat{z}}(z_q) &= \sup_{(p,a,b,e)} \min \mu_{\widehat{p}_{ijk}}(p_{ijk}), \mu_{\widehat{\mathscr{A}_i}}(a_i), \mu_{\widehat{\mathscr{B}}_j}(b_j), \mu_{\widehat{\mathscr{E}}_k}(e_k), \ \forall\ i, j, k \\ &= \mathscr{Z}(p, a, b, e),\end{aligned}$$

where $\mathscr{Z}(p, a, b, e)$ is defined in (4.1). The application of the extension principle to $\widehat{z}$ may be viewed as the application of this extension principle to the $\beta-$ cuts of $\widehat{z}$. Denote the $\beta-$ cuts of $\widehat{\mathscr{C}}_{ijk}$, $\widehat{\mathscr{A}_i}$, $\widehat{\mathscr{B}}_j$, and $\widehat{\mathscr{E}}_k$ as

$$\begin{aligned}(\mathscr{C}_{ijk})_\beta &= p_{ijk} \in \mathscr{S}(\widehat{\mathscr{C}}_{ijk})|\mu_{(\mathscr{C}_{ijk})}(p_{ijk}) \geq \beta = [(p_{ijk})^{\mathcal{L}}_\beta, (p_{ijk})^{U}_\beta] \\ (\mathscr{A}_i)_\beta &= a_i \in \mathscr{S}(\widehat{\mathscr{A}_i})|\mu_{(A_i)}(a_i) \geq \beta = [(a_i)^{\mathcal{L}}_\beta, (a_i)^{\mathcal{U}}_\beta] \\ (\mathscr{B}_j)_\beta &= b_j \in \mathscr{S}(\widehat{\mathscr{B}}_j)|\mu_{(B_j)}(b_j) \geq \beta = [(b_j)^{\mathcal{L}}_\beta, (b_j)^{\mathcal{U}}_\beta] \\ (\mathscr{E}_k)_\beta &= e_k \in \mathscr{S}(\widehat{\mathscr{E}}_k)|\mu_{(E_k)}(e_k) \geq \beta = [(e_k)^{\mathcal{L}}_\beta, (e_k)^{\mathcal{U}}_\beta].\end{aligned}$$

These gaps show where the amount of shipping price, supply, demand, and transportation are located at the β level of possibility. Lagrangian duality states that if one issue is unbounded, the other is impossible to address. Furthermore, if both issues are solvable, then neither of them have ideal solutions with an identical goal in mind. Alternatively, equation (4.12) can be reconstructed as

$$\begin{cases} \mathscr{Z}^{\mathcal{U}}_\beta = \text{maximize}_{(y,\lambda,\zeta)}\mathscr{Z} = -\frac{1}{2}\sum_{i=1}^{m}\sum_{j=1}^{n}\sum_{k=1}^{l} v_{ijk}y_{ijk} - \sum_{i=1}^{m} b_i\eta_i \\ \text{subject to} \\ \sum_{i=1}^{m}\sum_{k=1}^{l} v_{ijk}y_{ijk} + \sum_{i=1}^{m}\eta_i a_{ij} - \zeta_j = -\sum_{i=1}^{m} p_{ij}y_i \\ (\mathscr{C}_{ij})^{\mathcal{L}}_\beta \leq p_{ij} \leq (\mathscr{C}_{ij})^{\mathcal{U}}_\beta, \ \ i, j = 1, 2, \ldots, m, n \\ (\mathscr{A}_i)^{\mathcal{L}}_\beta \leq a_i \leq (\mathscr{A}_i)^{\mathcal{U}}_\beta, \ \ i = 1, 2, \ldots, m \\ (\mathscr{B}_j)^{\mathcal{L}}_\beta \leq b_j \leq (\mathscr{B}_j)^{\mathcal{U}}_\beta, \ \ j = 1, 2, \ldots, n \\ (\mathscr{E}_k)^{\mathcal{L}}_\beta \leq e_k \leq (\mathscr{E}_k)^{\mathcal{U}}_\beta, \ \ k = 1, 2, \ldots, l \\ \eta_i, \ \zeta_j, \ y_{ijk} \geq 0. \end{cases} \tag{4.13}$$

The generated target value is a global optimum solution that addresses the quadratic programming issue that this framework describes. The upper bound of the intended values for the quadratic fuzzy transportation issue indicates the optimal solution, $\mathscr{Z}^{\mathcal{U}}$.

Lower Bound: The linear programming duality theorem makes it apparent that the dual and primal models possess an identical objective value. The result of the Eq. (4.3) is

$$\begin{cases} \mathscr{Z}_v^{\mathcal{L}} = \text{minimize} \sum_{i=1}^{m}\sum_{j=1}^{n} \mathscr{C}_{ij} y_{ij} + \frac{1}{2}\sum_{i=1}^{m}\sum_{j=1}^{n}\sum_{k=1}^{l} v_{ijk} y_{ijk} \\ \text{subject to} \\ \sum_{j=1}^{n}\sum_{k=1}^{l} y_{ijk} = \mathscr{A}_i \\ \sum_{i=1}^{m}\sum_{k=1}^{l} y_{ijk} = \mathscr{B}_j \\ \sum_{i=1}^{m}\sum_{j=1}^{n} y_{ijk} = \mathscr{E}_k \\ (\mathscr{C}_{ij})_\beta^{\mathcal{L}} \le p_{ij} \le (\mathscr{C}_{ij})_\beta^{\mathcal{U}}, \quad i,\ j = 1, 2, \ldots, m, n \\ (\mathscr{A}_i)_\beta^{\mathcal{L}} \le a_i \le (\mathscr{A}_i)_\beta^{\mathcal{U}}, \quad i = 1, 2, \ldots, m \\ (\mathscr{B}_j)_\beta^{\mathcal{L}} \le b_j \le (\mathscr{B}_j)_\beta^{\mathcal{U}}, \quad j = 1, 2, \ldots, n \\ (\mathscr{E}_k)_\beta^{\mathcal{L}} \le e_k \le (\mathscr{E}_k)_\beta^{\mathcal{U}}, \quad k = 1, 2, \ldots, l \\ y_{ijk} \ge 0. \end{cases} \tag{4.14}$$

We have $(\mathscr{C}_{ij})_\beta^{\mathcal{L}}\, y_{ij} \le p_{ij} y_{ij} \le (\mathscr{C}_{ij})_\beta^{\mathcal{U}} y_{ij}$, which is in (4.2). The context of determining the objective function's minimal value:

$$\text{minimize } \mathscr{Z} = \sum_{i=1}^{m}\sum_{j=1}^{n} (\mathscr{C}_{ij})_\beta^{\mathcal{L}} y_{ij} + \frac{1}{2}\sum_{i=1}^{m}\sum_{j=1}^{n}\sum_{k=1}^{l} v_{ijk} y_{ijk}.$$

4.5.1 Quadratic Interval STP

The aforementioned quadratic interval solid transportation problem (QISTP) incorporates the following:

$$\begin{cases} \text{minimize } \ \mathscr{Z}^{v} = \sum_{i=1}^{m}\sum_{j=1}^{n}\sum_{k=1}^{l}[p^{q}_{\mathcal{L}ijk}, p^{q}_{\mathcal{R}ijk}] + \frac{1}{2}\sum_{i=1}^{m}\sum_{j=1}^{n}\sum_{k=1}^{l} v_{ijk}y_{ijk} \\ \text{subject to} \\ \sum_{j=1}^{n}\sum_{k=1}^{l} y_{ijk} = \mathscr{A}_i \\ \sum_{i=1}^{m}\sum_{k=1}^{l} y_{ijk} = \mathscr{B}_j \\ \sum_{i=1}^{m}\sum_{j=1}^{n} y_{ijk} = \mathscr{E}_k \\ y_{ijk} \geq 0, \quad \text{for all } i,\ j,\ k \end{cases} \tag{4.15}$$

where $\mathscr{A}_i = [a_{Li}, a_{Ri}],\ i = 1, 2, \ldots, m,\ \mathscr{B}_j = [b_{Lj}, b_{Rj}],\ j = 1, 2, \ldots, n,\ \mathscr{E}_k = [e_{Lk}, e_{Rk}],\ k = 1, 2, \ldots, l$ are intervals of possible values for the supplies, demands, and conveyance capacities. By stating the lower and upper bounds, the range of the target values in problem (4.3) is capable of being determined. The lower bound's simpler situation is covered initially. The principle of crisp formulation implies that a fuzzy constraint problem can be seen as an accumulation of crisp constraint problems, which is evidently true for equation (4.2). It is appropriate to refer to constraints of this type as predicates or sets of values for a fixed subset of variables. Consider the following multi-objective QISTP:

$$\begin{cases} \mathscr{Z}^{\mathcal{L}}_{v} = \text{minimize } \ \mathscr{Z} \\ = \sum_{i=1}^{m}\sum_{j=1}^{n}\sum_{k=1}^{l}[p^{v}_{\mathcal{L}ijk}, p^{v}_{\mathcal{R}ijk}] + \frac{1}{2}\sum_{i=1}^{m}\sum_{j=1}^{n}\sum_{k=1}^{l} v_{ijk}y_{ijk} \\ \text{subject to} \\ \sum_{j=1}^{n}\sum_{k=1}^{l} y_{ijk} \leq a_{\mathcal{R}i}; \qquad \sum_{j=1}^{n}\sum_{k=1}^{l} y_{ijk} \geq a_{\mathcal{L}i}; \\ \sum_{i=1}^{m}\sum_{k=1}^{l} y_{ijk} \leq b_{\mathcal{R}j}; \qquad \sum_{i=1}^{m}\sum_{k=1}^{l} y_{ijk} \geq b_{\mathcal{L}j}; \\ \sum_{i=1}^{m}\sum_{j=1}^{n} y_{ijk} \leq e_{\mathcal{R}k}; \qquad \sum_{i=1}^{m}\sum_{j=1}^{n} y_{ijk} \geq e_{\mathcal{L}k}; \\ \mathscr{C}^{\mathcal{L}}_{ij} \leq p_{ij} \leq \mathscr{C}^{\mathcal{R}}_{ij}, \quad j = 1, 2, \ldots, n, \ \ i = 1, 2, \ldots, m \\ \mathscr{A}^{\mathcal{L}}_{i} \leq a_i \leq \mathscr{A}^{\mathcal{R}}_{i}, \quad i = 1, 2, \ldots, m \\ \mathscr{B}^{\mathcal{L}}_{j} \leq b_j \leq \mathscr{B}^{\mathcal{R}}_{j}, \quad j = 1, 2, \ldots, n \\ \mathscr{E}^{\mathcal{L}}_{k} \leq e_k \leq \mathscr{E}^{\mathcal{R}}_{k}, \quad k = 1, 2, \ldots, l \\ v^{\mathcal{L}}_{ijk} \leq v_{ijk} \leq v^{\mathcal{R}}_{ijk} \\ y_{ijk} \geq 0. \end{cases} \tag{4.16}$$

Example 4.1 Consider the following quadratic fuzzy transportation problem:

$$
\begin{cases}
\text{minimize } \mathscr{Z} = (-10,-6)y_{111} + (-5,-3)y_{222} + (1,2)y_{333} + 4y_{111}^2 \\
\quad +4y_{222}^2 + 4y_{333}^2 + 8y_{111}y_{222} + 8y_{222}y_{333} + 8y_{333}y_{111} \\
\text{subject to} \\
(2,4)y_{111} + 6y_{222} + (1,2)y_{333} \leq (2,20) \\
(4,6)y_{111} + (-1,-0.5)y_{222} + y_{333} \leq (4,6) \\
(-2,8)y_{111} + (1,4)y_{222} + (3,2)y_{333} \leq (1,2) \\
\sum_{j=1}^{3}\sum_{k=1}^{2} y_{3jk} = [17,21] \\
\sum_{j=1}^{3}\sum_{k=1}^{2} y_{1jk} = [24,8] \\
\sum_{j=1}^{3}\sum_{k=1}^{2} y_{2jk} = [18,19] \\
\sum_{i=1}^{3}\sum_{k=1}^{2} y_{i2k} = [15,17] \\
\sum_{i=1}^{3}\sum_{k=1}^{2} y_{i1k} = [2,4] \\
\sum_{i=1}^{3}\sum_{k=1}^{2} y_{i3k} = [10,12] \\
y_{111}, y_{222}, y_{333} \geq 0.
\end{cases}
$$

The upper bound of the objective value $Z_\beta^{\mathcal{U}}$ by (4.1) can be formulated as

$$
\begin{cases}
\mathscr{Z}_\beta^{\mathcal{U}} = \text{maximize } -\frac{1}{2}(8y_{111}^2 + 8y_{222}^2 + 8y_{333}^2 + 16y_{111}y_{222} \\
+16y_{222}y_{333} + 16y_{333}y_{111}) - 2\lambda_1 - 4\lambda_2 - \lambda_3 \\
\text{subject to} \\
8y_{111} - 4y_{222} + y_{333} + r_{111} + r_{211} + r_{311} - \Delta_1 = 6 \\
-4y_{111} + 2y_{222} - 4y_{333} + r_{121} + r_{221} + r_{321} - \Delta_2 = 3 \\
2y_{111} - 4y_{222} + 2y_{333} + r_{123} + r_{223} + r_{323} - \Delta_3 = -2 \\
-10 \leq C_1 \leq -6 \\
-5 \leq C_2 \leq -3 \\
1 \leq C_3 \leq 2 \\
2\lambda_1 \leq r_{111} \leq 4\lambda_1 \\
4\lambda_2 \leq r_{211} \leq 6\lambda_2 \\
-2\lambda_3 \leq r_{311} \leq 8\lambda_3 \\
6\lambda_1 = r_{121} \\
-\lambda_2 \leq r_{222} \leq 0.5\lambda_2 \\
\lambda_3 \leq r_{223} \leq 4\lambda_3 \\
\lambda_1 \leq r_{331} \leq 2\lambda_1 \\
\lambda_2 = r_{332} \\
3\lambda_3 \leq r_{333} \leq 2\lambda_3 \\
\lambda_1, \lambda_2, \lambda_3, \Delta_1, \Delta_2, \Delta_3 \geq 0.
\end{cases}
$$

By using (Lau et al. 2009), we obtain the stationary point $Z_\beta^{\mathcal{U}} = -277$, which occurs at

$$\lambda_1 = 2,\ \ \lambda_2 = 4,\ \ \lambda_3 = 1,\ \ r_{311} = 6,\ \ r_{321} = 3,\ \ \Delta_1 = \Delta_2 = 0,\ \ \Delta_3 = 2,$$
$$\text{and } y_{312} = 1,\ \ y_{112} = 1,\ \ y_{122} = 7,\ \ y_{222} = 8,\ \ y_{232} = 10,\ \ p_3 = 1.$$

The lower bound of the objective value $\mathscr{Z}_\beta^{\mathcal{L}}$ according to model (4.2) can be formulated as

$$Z_\beta^{\mathcal{L}} = \text{minimize} \ -10y_{111} - 5y_{222} + y_{333} + \frac{1}{2}[\,8y_{111}^2 + 8y_{222}^2 + 8y_{333}^2 + 16y_{111}y_{222} + 16y_{222}y_{333} + 16y_{333}y_{111}\,]$$

$$\begin{cases} \text{subject to} \\ \sum_{j=1}^{3}\sum_{k=1}^{2} y_{3jk} \le 17; & \sum_{j=1}^{3}\sum_{k=1}^{2} y_{3jk} \ge 21; \\ \sum_{j=1}^{3}\sum_{k=1}^{2} y_{1jk} \le 24; & \sum_{j=1}^{3}\sum_{k=1}^{2} y_{1jk} \ge 8; \\ \sum_{j=1}^{3}\sum_{k=1}^{2} y_{2jk} \le 18; & \sum_{j=1}^{3}\sum_{k=1}^{2} y_{2jk} \ge 19; \\ \sum_{i=1}^{3}\sum_{k=1}^{2} y_{i2k} \le 15; & \sum_{i=1}^{3}\sum_{k=1}^{2} y_{i2k} \ge 17; \\ \sum_{i=1}^{3}\sum_{k=1}^{3} y_{i1k} \le 2; & \sum_{i=1}^{3}\sum_{k=1}^{2} y_{i1k} \ge 4; \\ \sum_{i=1}^{3}\sum_{k=1}^{2} y_{i3k} \le 10; & \sum_{i=1}^{3}\sum_{k=1}^{2} y_{i3k} \ge 12; \\ y_{ijk} \ge 0. \end{cases}$$

The global optimum solution $\mathscr{Z}_\beta^{\mathcal{L}} = -6.250$, $y_{111} = 1.250$, $y_{222} = 0$, and $y_{333} = 0$. By combining these two findings, one can arrive at the conclusion that the objective values of this interval linear program fall between $[-6.250, -277]$.

4.6 Conclusion

Transportation models are commonly used in supply chains and logistics to save costs. The solid transportation problem takes into account supply and demand as well as transportation capacity in order to meet transportation demands in a way that is both economical and effective. A multi-objective quadratic transportation problem was examined using the two-level mathematical programming paradigm.

The theoretical applications are presented and the optimal solution to the relative quadratic transportation problem is found. The recommended method may effectively determine the range of the objective values for the interval quadratic programming issue, as demonstrated by a practical example. Future research should concentrate on creating a fractional neural network and goal programming approach for engineering disciplines with numerous possible applications.

Conflict of Interest The authors declare no conflict of interest.

References

Abo-Sinna MA (2001) A bi-level non-linear multi-objective decision making under fuzziness. J Oper Res Soc 38:484–495

Akram M, Muhammad Umer Shah S, Ali Al-Shamiri M, Edalatpanah SA (2022) Fractional transportation problem under interval-valued Fermatean fuzzy sets. AIMS Math 7:17327–17348

Arora SR, Khurana A (2004) Three dimensional fixed charge bi-criterion indefinite quadratic transportation problem. Yugoslav J Oper Res 14:83–97

Bector CR, Chandra S, Dutta J (2005) Principles of optimization theory. Narosa Publishing House

Bellman RE, Zadeh LA (1970) Decision making in a fuzzy environment. Manag Sci 17:141–164

Bit AK, Biswal MP, Alam SS (1993) An additive fuzzy programming model for multi-objective transportation problem. Fuzzy Sets Syst 57:313–319

Buckly JJ, Feuring T (2000) Evolutionary algorithm solution to fuzzy problems: fuzzy linear programming. Fuzzy Sets Syst 109:35–53

Chanas S (1983) The use of parametric programming in fuzzy linear programming. Fuzzy Sets Syst 11:243–251

Chanas S, Kuchta D (1998) Fuzzy integer transportation problem. Fuzzy Sets Syst 98:291–298

Chanas S, Kolodziejczyk W, Machaj A (1984) A fuzzy approach to the transportation problem. Fuzzy Sets Syst 13:211–221

Charnes A, Cooper WW (1954) The stepping-stone method for explaining linear programming calculation in transportation problem. Manag Sci 1:49–99

Dahiya K, Verma V (2007) Capacitated transportation problem with bounds on rim conditions. Eur J Oper Res 178:718–737

Delgado M, Verdegay JL, Vila MA (1989) A general model for fuzzy linear programming. Fuzzy Sets Syst 29:21–29

Diaz JA (1979) Finding a complete description of all efficient solutions to a multi-objective transportation problem. Ekonomicko—Matematicki Obzor 15:62–73

George B, Thapa MN (1997) Linear programming introduction, Springer series in operations research and financial engineering

Gupta SK, Dangar D (2010) Duality in fuzzy quadratic programming with exponential membership functions. Fuzzy Inf Eng 4:337–346

Gupta P, Mehlawat MK (2007) An algorithm for a fuzzy transportation problem to select a new type of coal for a steel manufacturing unit. TOP 15:114–137

Haley KB (1962) The solid transportation problem. Oper Res 10:448–463

Haley KB (1963) The multi index problem. Oper Res 11:368–379

Hitchcock FL (1941) The distribution of a product from several sources to numerous localities. J Math Phys 20:224–230

Horcik R (2008) Solution of a system of linear equation with fuzzy numbers. Fuzzy Sets Syst 159:1788–1810

Ida K, Gen M, Li Y (1996) Neural networks for solving multi criteria solid transportation problem. Comput Ind Eng 31:873–877

Jain M, Saksena PK (2012) Time minimizing transportation problem with fractional bottleneck objective function. Yugoslav J Oper Res 22:115–129

Kumar A, Kaur A (2011) Application of classical transportation methods to find the fuzzy optimal solution of fuzzy transportation problem. Fuzzy Inf Eng 1:81–99

Kumar A, Kaur A (2012) Methods for solving unbalanced fuzzy transportation problems. Int J Oper Res 12:287–316

Kumar A, Kaur A, Gupta A (2011) Fuzzy linear programming approach to solving fuzzy transportation problems with transhipment. J Math Model Algorithm 10:163–180

Lau HCW, Chan TM, Tsui WT, Chan FTS, Ho GTS, Choy KL (2009) A fuzzy guided multi-objective evolutionary algorithm model for solving transportation problem. Expert Syst Appl 36:8255–8268

Li W, Tian X (2008) Numerical solution method for general interval quadratic programming. Appl Math Comput 202:589–595

Li Y, Ida K, Gen M (1997) Improved genetic algorithm for solving multi-objective solid transportation problem with fuzzy numbers. Comput Ind Eng 33:589–592

Liu ST (2009) Quadratic programming with fuzzy parameters: a membership function approach. Chaos Solitons Fractals 40:237–245

Liu ST, Wang RT (2007) A numerical solution method to interval quadratic programming. Appl Math Comput 189:1274–1281

Mahdavi-Amiri N, Nasseri SH (2007) Duality results and a dual simplex method for linear programming problems with trapezoidal fuzzy variables. Fuzzy Sets Syst 17:1961–1978

Marbini AH, Tavana M (2011) An extension of the linear programming method with fuzzy parameters. Int J Math Oper Res 3:44–55

Muhammada A, Shah U, Muhammada S, Allahviranloo T (2023) A new method to determine the Fermatean fuzzy optimal solution of transportation problems. J Intell Fuzzy Syst 44:309–328

Nehi HM, Maleki HR, Mashinchi M (2006) A canonical representation for the solution of fuzzy linear systems and fuzzy linear programming problem. J Appl Math Comput 20:345–354

Oh Eigeartaigh M (1982) A fuzzy transportation algorithm. Fuzzy Sets Syst 8:235–243

Ozkok BA, Tiryaki F (2011) A compensatory fuzzy approach to multi-objective linear supplier selection problem with multiple-item. Expert Syst Appl 38:11363–11368

Pal BB, Moitra BN, Sen S (2011) A linear Goal Programming approach to multi-objective fractional programming with interval parameter sets. Int J Math Oper Res 3:697–714

Patel G, Tripathy J (1989) The solid transportation problem and its variants. Int J Manag Syst 5:17–36

Poonam S, Abbas SH, Gupta VK (2012) Fuzzy transportation problem of triangular numbers with β-cut and ranking technique. J Eng 2:1162–1164

Pramanik S, Roy TK (2008) Multi objective transportation model with fuzzy parameters: priority based fuzzy goal programming approach. J Transp Syst Eng Inf Technol 8:40–48

Preetvanti S, Saxena PK (2003) The multi objective time transportation problem with additional restrictions. Eur J Oper Res 146:460–476

Punnen AP, Zhang R (2011) Quadratic bottleneck problems. Naval Res Logistics 58:153–164

Qiao Z, Zhang Y, Wang G (1994) On fuzzy random linear programming. Fuzzy Sets Syst 65:31–49

Sahoo L (2023) Transportation problem in fuzzy environment. RAIRO-Oper Res 57:145–156

Tada M, Ishii H (1996) An integer fuzzy transportation problem. Comput Math Appl 31:71–87

Wu XL, Liu YK (2012) Optimizing fuzzy portfolio selection problems by parametric quadratic programming. Fuzzy Optim Decision Making 11:411–449

Wu H, Zhang K (2008) A new accelerating method for global non-convex quadratic optimization with non-convex quadratic constraints. Appl Math Comput 197:810–818

Xie F, Jia R (2012) Nonlinear fixed charge transportation problem by minimum cost flow-based genetic algorithm. Comput Ind Eng 63:763–778

Yager RR (2008) Level set and the extension principle for interval valued fuzzy sets and its application to uncertainty measures. Inf Sci 178:3565–3576

Yang L, Liu L (2007) Fuzzy fixed charge solid transportation problem and algorithm. Appl Soft Comput 7:879–889

Zadeh LA (1965) Fuzzy sets. Inf Control 8:338–353

Zadeh LA (1978) Fuzzy sets as a basis for a theory of possibility. Fuzzy Sets Syst 1:3–28

Zimmermann HJ (1987) Fuzzy sets decision—making and expert systems. Kluwar Academic Publisher, Boston

Zimmermann HJ (1992) Fuzzy mathematical programming. In: Stauart C, Shapiroed (ed) Encyclopedia of artificial intelligence, vol 1. Wiley, pp 521–528

Zimmermann HJ (1976) Description and Optimization to fuzzy systems. Int J General Syst 2:209–215

Zimmermann HJ (1978) Fuzzy programming and linear programming with several objective functions. Fuzzy Sets Syst 1(1):45–55

Chapter 5
A Study on Fuzzy Set and Its Extensions

Arpita Bhowmik, Madhumangal Pal, Laxminarayan Sahoo, and Sovan Samanta

Abstract In our real life problems many ambiguities and uncertainties exist. So it is difficult to express such problems in terms of classical set theory which only considers two phases 'belongs to' and 'not belongs to'. So classical set theory is unable to capture such uncertainties. In 1965, Lotfi Asker Zadeh first introduced the notion of a fuzzy set. The fuzzy set is able to capture uncertainties in our real-life problems. Later many extensions of fuzzy sets are introduced. The natural question arises about the motivation and relations among all the extensions of fuzzy sets. In this study, the fuzzy set and its extensions such as type-2 fuzzy set, interval-valued fuzzy set, intuitionistic fuzzy set, bipolar fuzzy set, neutrosophic fuzzy set, picture fuzzy set, rough set, and soft set are discussed. This study reviews recent developments of fuzzy sets and analyses their relations. Also, the motivation and application areas of all such extensions are studied here.

Keywords Fuzzy set · Type-2 fuzzy set · Interval-valued fuzzy set · Intuitionistic fuzzy set · Bipolar valued fuzzy set · Neutrosophic set · Picture fuzzy set · Rough set · Soft set

A. Bhowmik · M. Pal
Department of Applied Mathematics, Vidyasagar University, Midnapore 721102, India

L. Sahoo (✉)
Department of Computer and Information Science, Raiganj University, Raiganj 733134, India
e-mail: lxsahoo@gmail.com

S. Samanta
Department of Technical Sciences, Western Caspian University, Baku 1001, Azerbaijan

Department of Mathematics, Tamralipta Mahavidyalaya, Tamluk, West Bengal 721636, India
e-mail: sovan.samanta@wcu.edu.az

R. Arora and S. Arora (eds.), *Optimization Techniques for Sustainable Environment Under Uncertainty*, Engineering Optimization: Methods and Applications,
https://doi.org/10.1007/978-981-96-0389-3_5

5.1 Introduction

In our world, problems involve uncertainty and impreciseness. Before 1965, people believed that the probability theory could tackle different uncertainties. But in 1965, Professor Lotfi Asker Zadeh (1965) of the University of California claimed that many types of uncertainties could not be handled with the principle of probability theory, which works based on a crisp set or classical set. Professor Zadeh realized that something new be implemented to face real-world problems. A fuzzy set (FS) is a collection of entities with vague memberships. A fuzzy set representation is more acceptable than the classical binary representation. After Zadeh, many objections arrived with the advent of FS (Zadeh 2008). Some scientists created many objections to the thought of Professor Zadeh. But in 1970, some researchers in Japan studied FS theory. In 1970 Mamdani developed a fuzzy logic controller. In 1977, fuzzy logic (Zadeh 1996) was applied in some industrial areas in Japan and Europe. A celebrated application of fuzzy logic was in Sendai, Japan, in 1988, the Sendai subway system. Using fuzzy logic controller, the Sendai subway system became the world's smoothest subway system. A fuzzy logic system has also been used in heating, ventilation and air conditioning, control of the speed of a car etc. (Kahraman et al. 2016).

After Zadeh's introduction of fuzzy set (FS) or type-1 fuzzy set to handle uncertainties, researchers found many types of uncertainties could not be dealt with type-1 FS as its membership value is a real number between 0 and 1. There are a few types of uncertainties (Mendel and John 2002) as the meaning of linguistic words such as 'good', 'tall', 'nice' has relative measures to different people. The parameter which is used to characterize type-1 FS may be uncertain. In 1975, Zadeh (Zadeh 1975) first introduced type-2 fuzzy set (type-2 FS), and these types of uncertainties can be handled with the type-2 FS as the membership function of this type is itself fuzzy (Mizumoto and Tanaka 1976). The membership function of a type-2 FS is of three dimensions, whereas the membership function is of two dimensions of a type-1 FS.

FS is used to deal with uncertainties in real life, and a real value in [0,1] defines membership of a set S in the universal set U. But the membership function is uncertain in real and can not be defined as a definite value. So interval-valued fuzzy set (IVFS) was presented independently in 1975 by Zadeh, Grattan-Guiness (1976), Jahn (1975), and Sambuc (1975) to handle the ambiguity of the value of membership. In an IVFS an interval $\left[\mu_S^L(p), \mu_S^R(p)\right]$ is used to represent the value of membership of a FS, where $\mu_S^L(p)$ is the lower bound and $\mu_S^R(p)$ is the upper bound of the membership function.

In FS and IVFS, only truth membership supporting systems are considered, but in some applications like information fusion, belief systems, and expert systems, falsity membership contradicting systems need to be considered. Then, Atanassov (1986), in 1983, came up with a solution, introducing an intuitionistic fuzzy set (IFS). Here both truth membership $\mu_S(p) \in [0,1]$ and falsity membership $\nu_S(p) \in [0, 1]$ are considered where the sum of truth and falsity membership is not more than 1.

Again there is a conflict in IFS as it can tackle incomplete data, not indeterminate information, and in this type of FS the indeterminacy is automatically $1 - \mu_S(p) - \nu_S(p)$, which is dependent on truth and falsity membership functions (Smarandache 2005). In 1995, Smarandache (1999) introduced the Neutrosophic Set (NS), which can tackle the indeterminacy of real-life problems.

FS is extended to BVFS where the range set is [−1,1] instead of [0,1]. It was introduced by Zhang (1994) in 1994.

There are four possibilities of answers: yes, no, abstain, and refusal. So the concept of neutrality exists in real life, which was ignored in the neutrosophic set. For example, in the result of an election program, there are four outcomes: vote for, abstain, vote against, and refuse to vote. Here abstain is that the voting paper refuses to both result 'vote for' and 'vote against' but participate in that vote, whereas 'refusal of voting' means unacceptable voting papers. In the medical diagnosis process, neutrality is considered (Dutta and Ganju 2018). For example, chest pain is irrelevant to the diagnosis of malaria. Cuong and Kreinovich (2013) presented a picture fuzzy set (PFS) to deal with neutrality.

Modeling, reasoning, or computing are dependent on tools that are crisp and so precise in character. There are a lot of complexities in medical science, economics, etc., that are not crisp in character. The theory of probability, FS, and IFS are used to tackle uncertainties, but there are some difficulties due to the inadequacy of parameters. Consequently, Molodsov (1999) introduced the soft set theory in 1999 as mathematical equipment to tackle uncertainties which is free from the above problems.

Pawlak first introduced rough set theory in 1982. Rough set theory is a new concept to dealing with imperfect data. It is an independent and complementary concept that is not competent with the theories that deal with imprecise data, such as fuzzy set theory. An equivalence relation is considered in rough set theory. Indiscernibility between the objects is the main concept here. Equivalence classes approximate crisp sets in rough set theory. Fuzzy set theory cannot deal with indiscernible relations, whereas membership function is irrelevant in rough set theory (Yao 1998; Pawlak 1985). Later few developments in fuzzy set theory are made in Samanta and Pal (2015), Samanta et al. (2016), Samanta et al. (2021).

5.2 Fuzzy Numbers

Fuzzy numbers are a more general form of binary numbers. It is useful in decision-making and data analysis.

Definition: A real fuzzy number n is any fuzzy subset of the real line whose membership function μ_n is.

(a) A continuous function from $\mathbb{R}$ to the closed interval [0,1].
(b) Constant on $[-\infty, p)$, $\mu_n(x) = 0, \forall x \in [-\infty, p)$
(c) Strictly increasing on $[p, q]$

(d) Strictly decreasing on $[q, r]$
(e) Constant on (r, ∞), $\mu_n(x) = 0, \forall x \in (r, \infty)$.

where p,q, and r are real number (Dubois and Prade 1978).

Definition: A triangular fuzzy number is a fuzzy number $n = (p, q, r)$, if its membership function μ_n is given by.

$$\mu_n(x) = \begin{cases} \frac{(x-p)}{(q-p)}, \forall x \in [p, q] \\ \frac{(r-x)}{(r-q)}, \forall x \in [q, r] \\ 0, \forall x \in (-\infty, p) \cup (r, \infty) \end{cases}$$

Definition: A fuzzy number $n = (p, q, r, s)$ is said to be a trapezoidal fuzzy number if its membership function is given as follows:

$$\mu_n(x) = \begin{cases} \frac{(x-p)}{(q-p)}, \forall x \in \left[p, q\right] \\ \mu_n(x) = 1, \forall x \in \left(q, r\right] \\ \frac{(r-x)}{(s-r)}, \forall x \in (r, s] \\ 0, \forall x \in (-\infty, p) \cup (s, \infty) \end{cases}$$

5.3 Fuzzy Set

Fuzzy set captures the membership of each element in [0,1]. The formal definition is given as follows.

Definition 1: A fuzzy set (FS) S in universal set U is a set of ordered pairs denoted as follows $S = \{(p, \mu_S(p))|p \in U\}$, where $\mu_S(p) : U \rightarrow [0,1]$ is called the membership function.

Note 1: Let S and T be two FSs defined on the same universal set U. Then the union of two FS S and T is denoted by $S \cup T$, shown in Fig. 5.1, and its membership values are obtained as follows:

$$\mu_{S \cup T}(p) = \max\{\mu_S(p), \mu_T(p)\}.$$

Note 2: Let S and T be two FSs defined on the same universal set U . The intersection of two FS is denoted by S $\cap T$, shown in Fig. 5.2, and the membership function is given by:

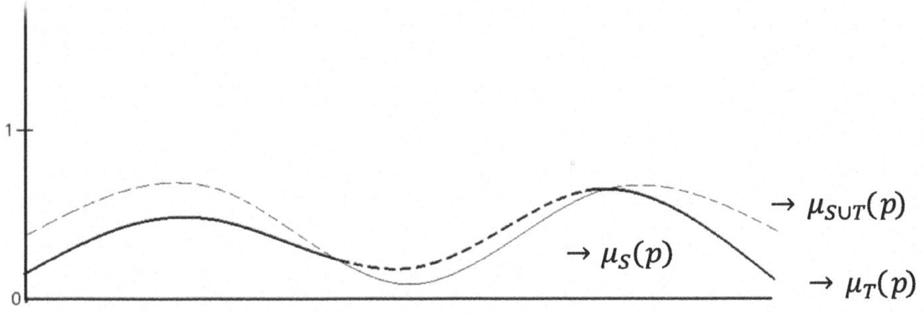

Fig. 5.1 Union of two FSs

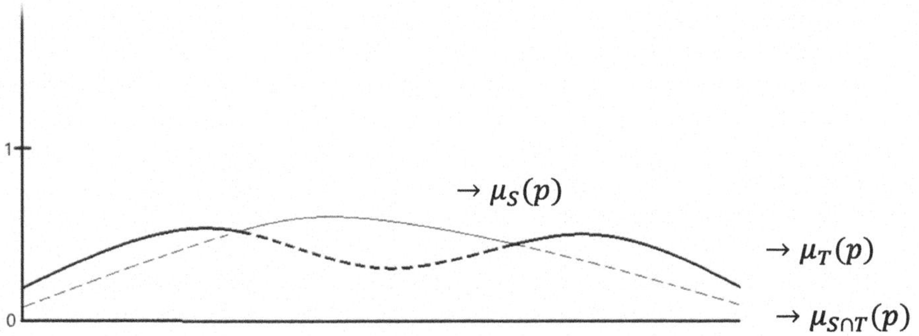

Fig. 5.2 Intersection of two FSs

$$\mu_{S\cap T}(p) = min\{\mu_S(p), \mu_T(p)\}$$

Note 3: Let S be a FS defined on universal set U. The complement of FS S is denoted by $\overline{S}$, shown in Fig. 5.3 and the membership function is given by

$$\mu_{\overline{S}}(p) = 1 - \mu_S(p), \forall p \in U$$

Type-2 fuzzy set- Let U be the universal set. A type-2 fuzzy set(type-2 FS) S is defined by

$$S = \{((p, u), \mu_S(p, u))|\forall p \in U, u \in J_p \subseteq [0,1]\},$$

where $\mu_S(p, u) \in [0,1]$ is called type-2 membership function (Castillo et al. 2007).

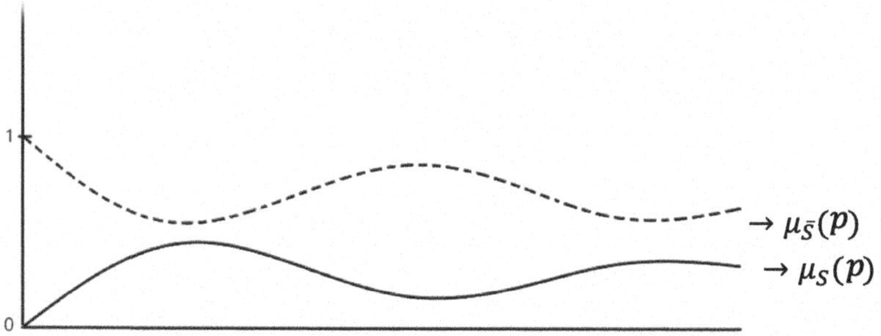

Fig. 5.3 Complement of a FS

Applications

(1) The fuzzy set theory is very useful for calculating solar cell parameters in indoor measurements. In the laboratory, some basic parameters of the silicon solar cell are practically measured. Fuzzy set theory is used to calculate PV parameters. The fuzzy set method is very helpful for comparing optimal values of the photovoltaic cell parameters (2021).
(2) An advanced cyber indicator based on the fuzzy rule is very beneficial for cyber security systems. A cyber security system can work flawlessly if the administrators are warned of future threats. This advanced cyber security system (ACSS) cannot prevent cyber attacks but warns of expected cyber threats (Alam and Pandey 2014).
(3) To study human behavior, fuzzy-rule-based analysis is very useful. It is also very helpful to the criminal investigation (Smithson and Oden 1999).
(4) Fuzzy graph theory is another branch of studying fuzzy sets. It has many applications in our real-life problems (Bhattacharya and Pal 2021, 2022).
(5) In social network analysis, fuzzy set theory especially fuzzy graph theory is very useful (Mahapatra et al. 2019, 2020).

5.4 Interval-Valued Fuzzy Set

Definition: Let U be the universal set. An interval-valued fuzzy set (IVFS) (Deschrijver and Cornelis 2007) S is defined as:

$$S = \left\{ \left(p, \left[\mu_S^L(p), \mu_S^R(p)\right]\right) | p \in U \right\}.$$

Note 1: Let S and T be two IVFSs defined on the same universal set U. The union of two IVFSs is done in Fig. 5.4 given by:

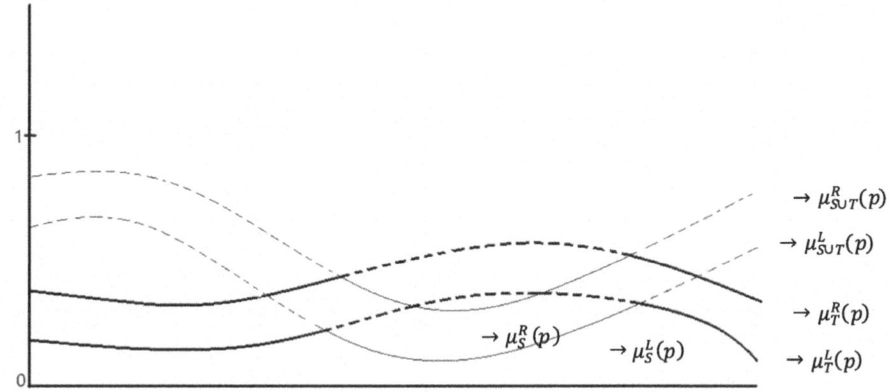

Fig. 5.4 Union of two IVFS

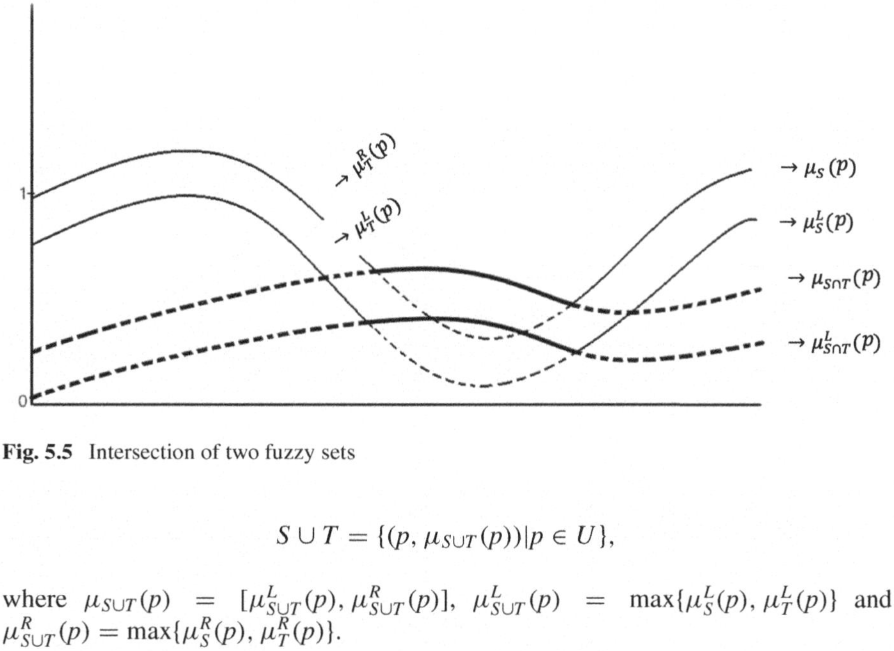

Fig. 5.5 Intersection of two fuzzy sets

$$S \cup T = \{(p, \mu_{S\cup T}(p))|p \in U\},$$

where $\mu_{S\cup T}(p) = [\mu^L_{S\cup T}(p), \mu^R_{S\cup T}(p)]$, $\mu^L_{S\cup T}(p) = \max\{\mu^L_S(p), \mu^L_T(p)\}$ and $\mu^R_{S\cup T}(p) = \max\{\mu^R_S(p), \mu^R_T(p)\}$.

Note 2: Let S and T be two IVFSs defined on the same universal set U. The intersection of two IVFSs is shown in Table 5.1 is given by:

$$S \cap T = \{(p, \mu_{S\cap T}(p))|p \in U\},$$

where $\mu_{S\cap T}(p) = [\mu^L_{S\cap T}(p), \mu^R_{S\cap T}(p)]$, $\mu^L_{S\cap T}(p) = \min\{\mu^L_S(p), \mu^L_T(p)\}$ and $\mu^R_{S\cap T}(p) = \min\{\mu^R_S(p), \mu^R_T(p)\}$.

Table 5.1 Fuzzy set

Year	Introducer	Types of fuzzy set
1965	L. A. Zadeh	Fuzzy set
1975	L. A. Zadeh	Type-2 fuzzy set
1975	Zadeh, Grattan-Guiness, Jahn, Sambuc	Interval-valued fuzzy set
1982	Pawlak	Rough set
1983	Atanassov	Intuitionistic fuzzy set
1994	Zhang	Bipolar-valued fuzzy set
1995	Smarandache	Neutrosophic Fuzzy set
1999	Molodsov	Soft set
2013	Cuong and Kreinovich	Picture fuzzy set
2014	Mesiarova-Zemankova	m-polar valued fuzzy set

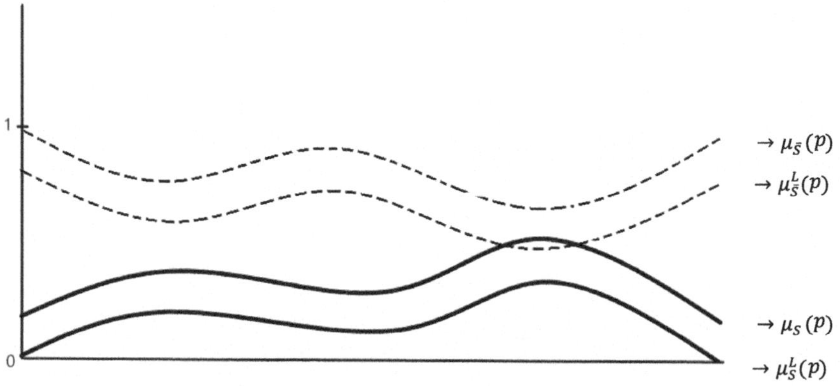

Fig. 5.6 Complement of an IVFS

Note 3: Let S be an IVFS defined on the universal set U. Then complement of IVFS is given as shown in Fig. 5.6.

$$\mu_{\overline{S}}(p) = \left[\mu_{\overline{S}}^{L}(p), \mu_{\overline{S}}^{R}(p)\right],$$

where $\mu_{\overline{S}}^{L}(p) = 1 - \mu_{S}^{R}(p)$ and $\mu_{\overline{S}}^{R}(p) = 1 - \mu_{S}^{L}(p)$.

5.5 Intuitionistic Fuzzy Set

Definition: Let U be the universal set. An intuitionistic fuzzy set (IFS) (Atanassov 1994) S in U is of the form.

$$S = \{(p, \mu_S(p), \nu_S(p)) | p \in U\},$$

where the functions $\mu_S(p) : U \to [0,1]$ and $\nu_S(p) : U \to [0,1]$ is the membership functions and non-membership functions of p, respectively. For any $p \in U$

$$0 \leq \mu_S(p) + \nu_S(p) \leq 1,$$

where $\pi_S(p) = 1 - \mu_S(p) - \nu_S(p)$ is called the intuitionistic index or hesitation part.

Note 1: Let S and T be two IFS defined on the same universal set U. Then the union of two IFS is given as shown in Fig. 5.7.

$$S \cup T = \{(p, \mu_{S \cup T}(p), \nu_{S \cup T}(p)) | p \in U\},$$

where $\mu_{S \cup T}(p) = \max\{\mu_S(p), \mu_T(p)\}$ and $\nu_{S \cup T}(p) = \min\{\nu_S(p), \nu_T(p)\}$.

Note 2: Let S and T be two IFS defined on the same universal set U. Then the intersection of two IFS is given as follows (see Fig. 5.8):

$$S \cap T = \{(p, \mu_{S \cap T}(p), \nu_{S \cap T}(p)) | p \in U\},$$

where $\mu_{S \cap T}(p) = \min\{\mu_S(p), \mu_T(p)\}$ and $\nu_{S \cap T}(p) = \max\{\nu_S(p), \nu_T(p)\}$.

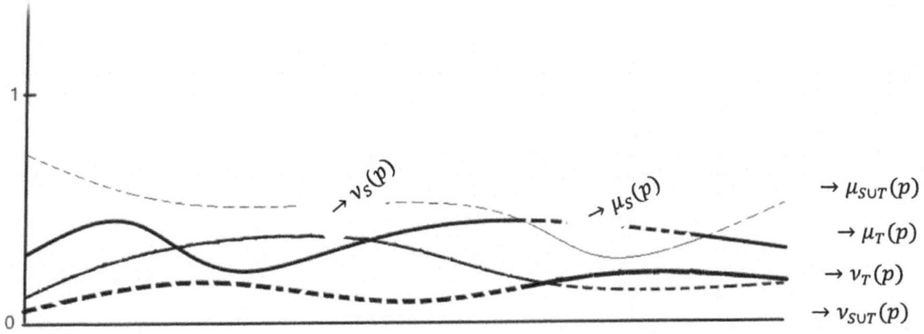

Fig. 5.7 Union of two IFS

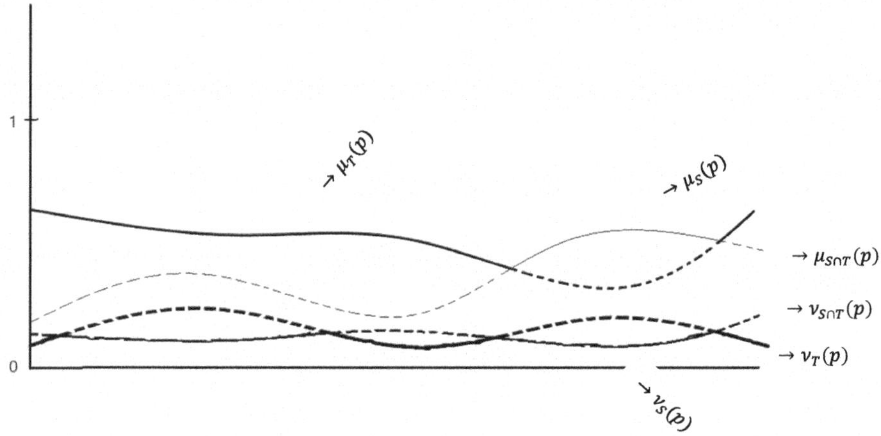

Fig. 5.8 Intersection of two IFS

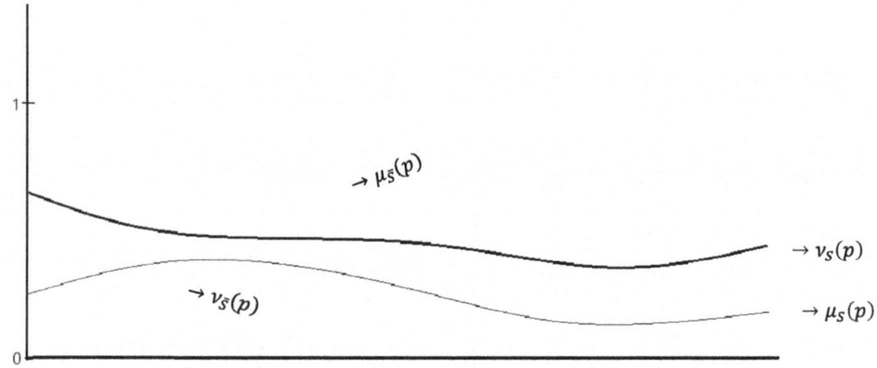

Fig. 5.9 Complement of an IFS

Note 3: Let S be an IFS in a universal set U. The complement of IFS is defined in Fig. 5.9 and defined as follows:

$$\overline{S} = \{(p, \mu_{\overline{S}}(p), \nu_{\overline{S}}(p)) | p \in U\},$$

where $\mu_{\overline{S}}(p) = \nu_S(p)$ and $\nu_{\overline{S}}(p) = \mu_S(p)$.

Applications

There are many applications of intuitionistic fuzzy set in medical diagnosis system and decision making process. In (Szmidt and Kacprzyk 2001) authors proposed a medical diagnosis system by introducing a distance function. The method is described as follows.

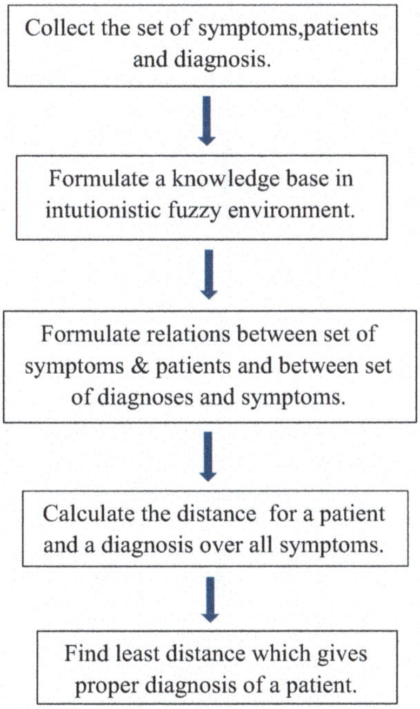

(1) In (Xu and Yager 2010) authors developed bonferroni aggregation operator in intuitionistic fuzzy environment and then applied it in an multicriteria decision making process. The method is as follows.

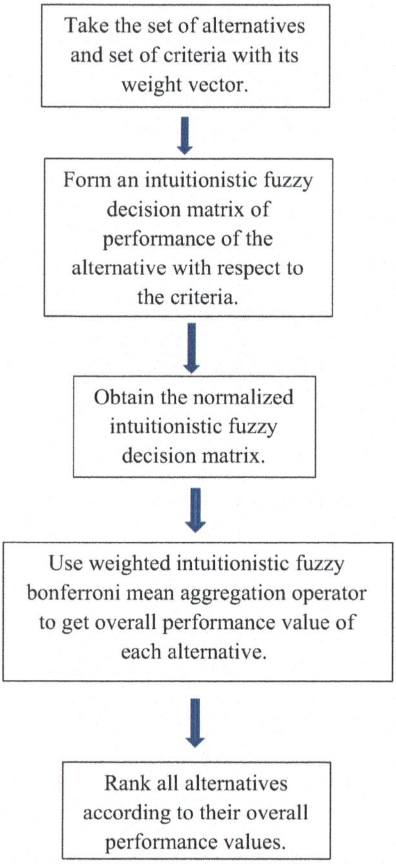

5.6 Bipolar-Valued Fuzzy Set

Definition: There are two types of representation of bipolar valued fuzzy set (BVFS) i. canonical representation ii. reduced representation (Lee 2004).

If U is the universal set, then the canonical representation of bipolar valued fuzzy set S in U is of the form:

$$S = \left\{ \left(p, \mu_S^P(p), \mu_S^N(p)\right) | p \in U \right\},$$

where $\mu_S^P(p) : U \rightarrow [0,1]$ is the positive membership degree, which denotes the degree of satisfaction of an element p to a property of a bipolar valued fuzzy set S and $\mu_S^N(p) : U \rightarrow [0,1]$ is the negative membership function, which is the satisfaction degree of p to some implicit counter property of S.

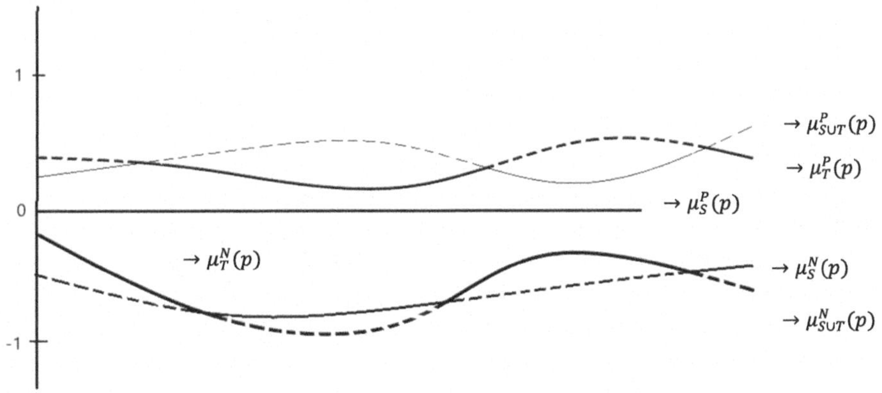

Fig. 5.10 Union of two BVFS

The reduced representation of a bipolar valued fuzzy set S in U is of the form:

$$S = \{(p, \mu^R_S(p))|p \in U\},$$

where $\mu^R_S(p) : U \to [-1,1]$.

Note 1: Let S and T be two BVFSs defined on the universal set U. The union (Lee 2000) of two BVFS is shown in Fig. 5.10 and defined as follows:

$$S \cup T = \{(p, (\mu^P_{S\cup T}(p), \mu^N_{S\cup T}(p)))|p \in U\},$$

where $\mu^P_{S\cup T}(p) = \max\{\mu^P_S(p), \mu^P_T(p)\}$ and $\mu^N_{S\cup T}(p) = \min\{\mu^N_S(p), \mu^N_T(p)\}$.

Note 2: Let S and T be two BVFSs defined on the universal set. Then the intersection of two BVFS is shown in Fig. 5.11 and defined as follows:

$$S \cap T = \{(p, (\mu^P_{S\cap T}(p), \mu^N_{S\cap T}(p)))|p \in U\},$$

where $\mu^P_{S\cap T}(p) = \min\{\mu^P_S(p), \mu^P_T(p)\}$ and $\mu^N_{S\cap T}(p) = \max\{\mu^N_S(p), \mu^N_T(p)\}$.

Note 3: Let S be a BVFSs defined on the universal set U. The complement of BVFS is shown in Fig. 5.12 and defined as follows:

$$\overline{S} = \{p, (\mu^P_{\overline{S}}(p), \mu^N_{\overline{S}}(p))|p \in U\},$$

where $\mu^P_{\overline{S}}(p) = 1 - \mu^P_S(p)$ and $\mu^N_{\overline{S}}(p) = -1 - \mu^N_S(p)$.

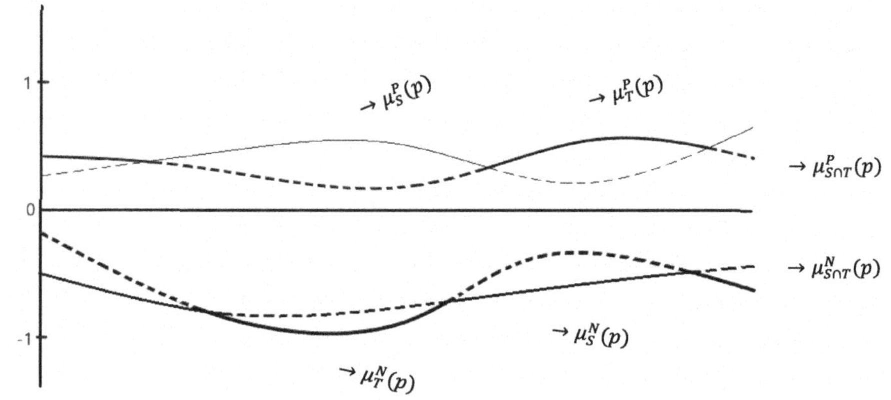

Fig. 5.11 Intersection of two BVFS

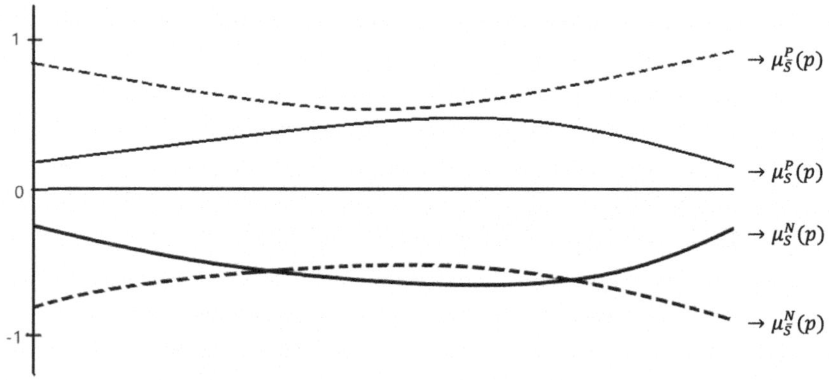

Fig. 5.12 Complement of a BVFS

Applications

There are many applications of the bipolar fuzzy set such as in the multi-attribute decision-making process. Here are some applications that have been discussed such as:

(1) In (Gulistan et al. 2021) complex bipolar fuzzy set is studied. A distance function and an aggregation operator in a complex bipolar fuzzy environment are discussed. Then the theory is applied to real-life problems. The method is as follows.

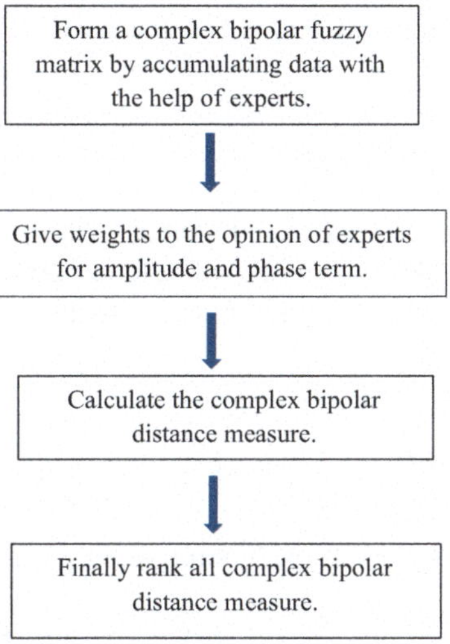

(2) In (Jana et al. 2020) authors introduced bipolar fuzzy Dombi prioritized averaging operator, bipolar fuzzy Dombi prioritized weighted geometric operator and applied these to multi attribute decision making problem. The method is as follows.

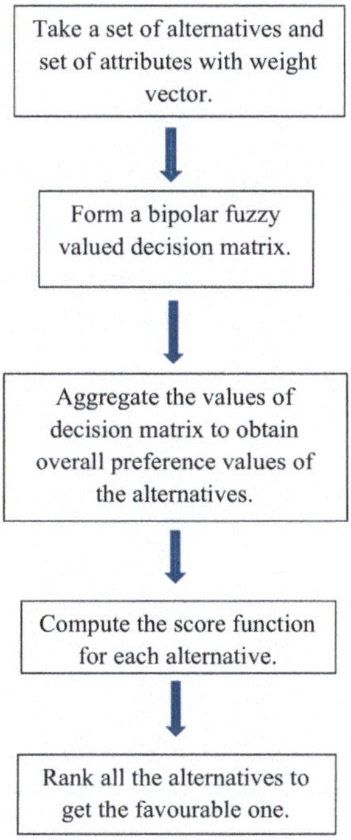

***m*-polar fuzzy set**- An m -polar fuzzy set (Chen et al. 2014) S of universal set U is a mapping from U to $[0,1]^m$.

BVFS is a particular type of m -polar fuzzy set, where $m = 2$.

Applications

m-polar fuzzy graph theory is studied in Pal et al. (2020) for the concept of domination. Also in Pramanik et al. (2016) interval-valued fuzzy set and the m-polar fuzzy set have been combined and are applied in real-life decision-making problems.

5.7 Neutrosophic Set

Definition: Let U be the universal set. A neutrosophic set (NS) S in U is of the form:

$$S = \{(p, T_S(p), F_S(p), I_S(p)) | p \in U\},$$

where $T_S(p) : U \to]0^-, 1^+[$ is the truth membership function, $F_S(p) : U \to]0^-, 1^+[$ is the falsity membership function and $I_S(p) : U \to]0^-, 1^+[$ is indeterminacy function such that $0^- \le supT_S(p) + supF_S(p) + supI_S(p) \le 3^+$.

Note 1: Let U be the universal set. Let S and T be two NSs defined on U. The union is shown in Figs. 5.13, 5.14 and 5.15 and is defined as follows:

$$S \cup T = \{(p, T_{S \cup T}(p), F_{S \cup T}(p), I_{S \cup T}(p)) | p \in U\},$$

where $T_{S \cup T}(p) = \max\{T_S(p), T_T(p)\}$, $F_{S \cup T}(p) = \min\{F_S(p), F_T(p)\}$, $I_{S \cup T}(p) = max\{I_S(p), I_T(p)\}$.

Note 2: Let S and T be two NSs (Wang et al. 2010) defined on universal set U. The intersection of two sets is shown in Figs. 5.16, 5.17, and 5.18 and defined as follows:

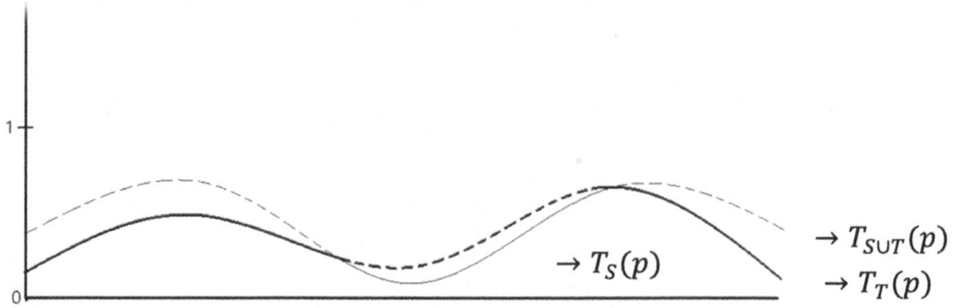

Fig. 5.13 Union of truth membership functions of two NSs

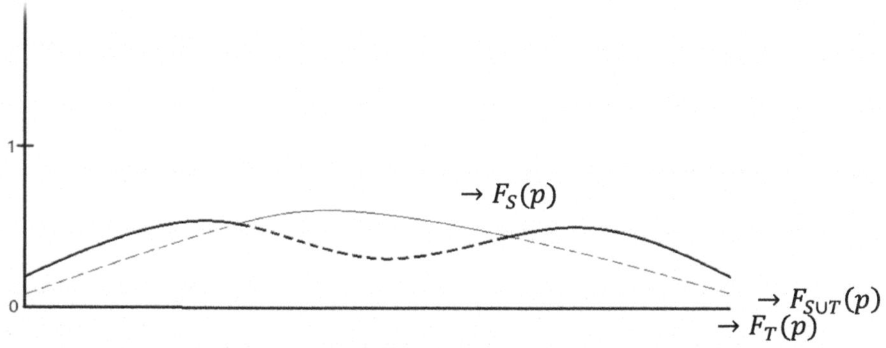

Fig. 5.14 Union of falsity membership functions of two NSs

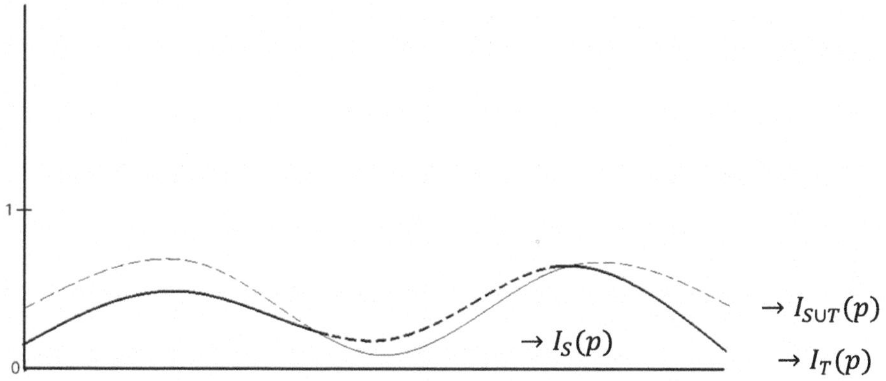

Fig. 5.15 Union of indeterminacy functions of two NSs

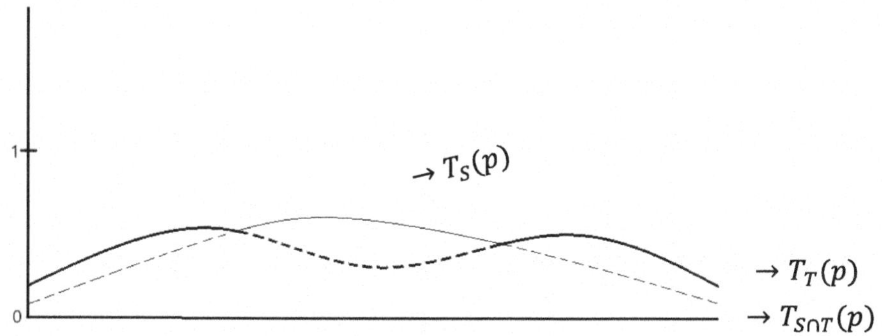

Fig. 5.16 Intersection of truth membership functions of two NSs

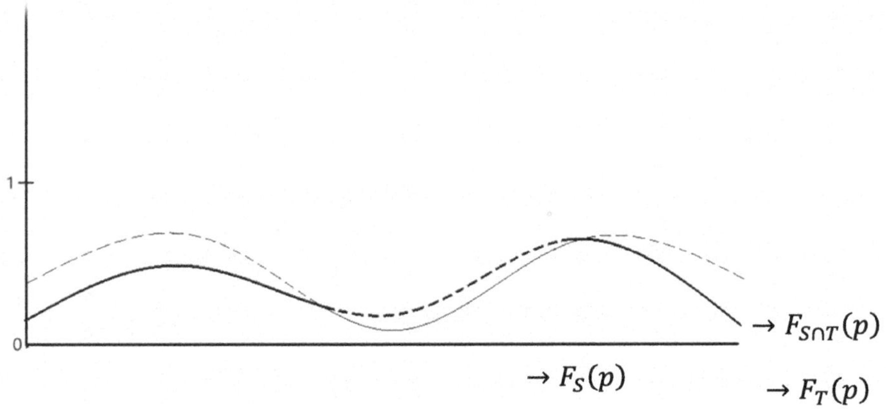

Fig. 5.17 Intersection of falsity membership functions of two NSs

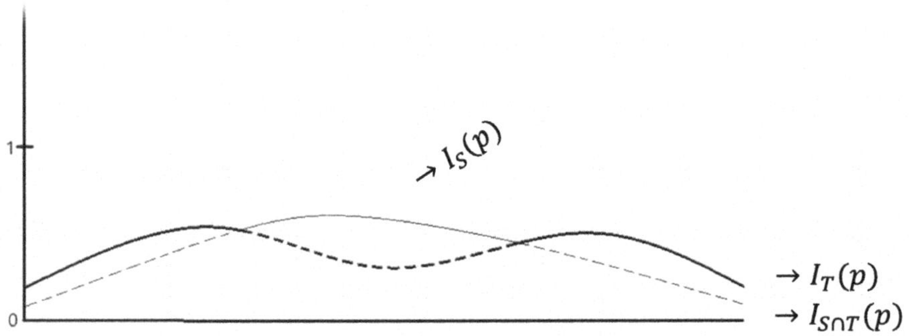

Fig. 5.18 Intersection of indeterminacy functions of two NSs

$$S \cap T = \{(p, T_{S\cap T}(p), F_{S\cap T}(p), I_{S\cap T}(p))|p \in U\},$$

where $T_{S\cap T}(p) = \min\{T_S(p), T_T(p)\}$, $F_{S\cap T}(p) = \max\{F_S(p), F_T(p)\}$ and $I_{S\cap T}(p) = min\{I_S(p), I_T(p)\}$

Note 3: Let S be a NSs defined on the universal set U. Then complement of S is shown in Fig. 5.19 and defined by

$$\overline{S} = \left\{\left(p, T_{\overline{S}}(p), F_{\overline{S}}(p), I_{\overline{S}}(p)\right)\right\},$$

where $T_{\overline{S}}(p) = F_S(p)$, $F_{\overline{S}}(p) = T_S(p)$ and $I_{\overline{S}}(p) = 1 - I_S(p)$.

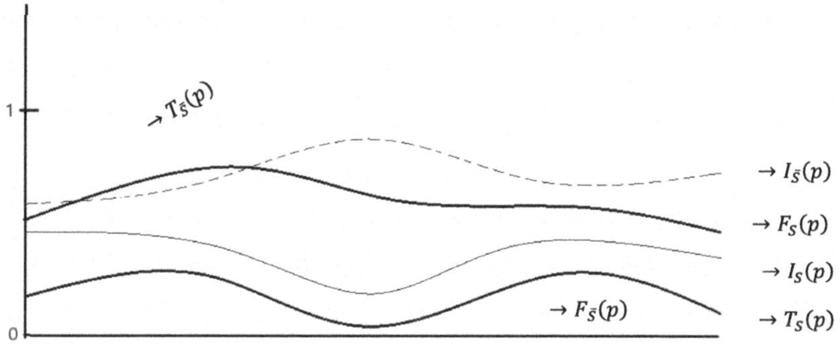

Fig. 5.19 Complement of a NS

Applications

A neutrosophic set is used to handle inconsistent and indeterminate information where neutrality is independent of truth and falsity membership function. A neutrosophic fuzzy set is studied in decision-making problems. Here one such example is discussed as follows.

In (Roan and Smarandache 2021) a new distance measure is introduced and then it is applied in a medical diagnosis process. The method is as follows.

5.8 Picture Fuzzy Set

Definition: A picture fuzzy set (PFS) (Cuong and Kreinovich 2014) S on a universal set U is of the form:

$$S = \left\{\left(p, \mu_S^P(p), \mu_S^I(p), \mu_S^N(p)\right) | p \in U\right\},$$

where $\mu_S^P(p) \in [0,1]$, $\mu_S^I(p) \in [0,1]$ and $\mu_S^N(p) \in [0,1]$ is called the degree of positive membership, degree of neutral membership and degree of negative membership of p in S. For any p in S, $\mu_S^P(p)$, $\mu_S^I(p) and \mu_S^N(p)$ satisfies that$\mu_S^P(p)+\mu_S^I(p)+\mu_S^N(p) \leq 1$.

For any for every element p in A, 1-($\mu_S^P(p) + \mu_S^I(p) + \mu_S^N(p)$) is called the degree of refusal membership of p in S.

Note 1: Let S and T be two PFSs defined on the universal set U. The union of two sets is defined as follows:

$$S \cup T = \left\{\left(p, \mu_{S\cup T}^P(p), \mu_{S\cup T}^I(p), \mu_{S\cup T}^N(p)\right) | p \in U\right\},$$

where $\mu_{S\cup T}^P(p) = \max\{\mu_S^P(p), \mu_T^P(p)\}$, $\mu_{S\cup T}^I(p) = \min\{\mu_S^I(p), \mu_T^I(p)\}$ and $\mu_{S\cup T}^N(p) = \min\{\mu_S^N(p), \mu_T^N(p)\}$

Note 2: Let S and T be two PFSs defined on the universal set U. Then the intersection of two sets is defined as follows:

$$S \cap T = \{(p, \mu^P_{S\cap T}(p), \mu^I_{S\cap T}(p), \mu^N_{S\cap T}(p))|p \in U\},$$

where $\mu^P_{S\cap T}(p) = \min\{\mu^P_S(p), \mu^P_T(p)\}$, $\mu^I_{S\cap T}(p) = \min\{\mu^I_S(p), \mu^I_T(p)\}$ and $\mu^N_{S\cap T}(p) = \max\{\mu^N_S(p), \mu^N_T(p)\}$

Note 3: Let S be PFS defined on the universal set U. Then complement of S is defined as follows:

$$\overline{S} = \{(p, \mu^P_{\overline{S}}(p), \mu^I_{\overline{S}}(p), \mu^N_{\overline{S}}(p))|p \in U\},$$

where $\mu^P_{\overline{S}}(p) = \mu^N_S(p)$, $\mu^I_{\overline{S}}(p) = \mu^I_S(p)$ and $\mu^N_{\overline{S}}(p) = \mu^P_S(p)$.

Note 4: Let S and T be two PFS on the universal set U, where S and T are defined as follows:

$$S = \{(x, \mu^P_S(x), \mu^I_S(x), \mu^N_S(x))|x \in U\}$$

$$T = \{(y, \mu^P_T(y), \mu^I_T(y), \mu^N_T(y))|y \in U\}$$

Then the addition of two sets is defined as follows:

$$S + T = \{(p, \mu^P_{S+T}(p), \mu^I_{S+T}(p), \mu^N_{S+T}(p))|p \in U\}$$

where $\mu^P_{S+T}(p) = \vee[\mu^P_S(x) \wedge \mu^P_T(y)]$, $\mu^I_{S+T}(p) = \wedge[\mu^I_S(x) \vee \mu^I_T(y)]$, $\mu^N_{S+T}(p) = \wedge[\mu^N_S(x) \vee \mu^N_T(y)]$ and $p = x + y$.

Note 4: Let S and T be two PFS on the universal set U, where S and T are defined as follows:

$$S = \{(x, \mu^P_S(x), \mu^I_S(x), \mu^N_S(x))|x \in U\}$$

$$T = \{(y, \mu^P_T(y), \mu^I_T(y), \mu^N_T(y))|y \in U\}$$

Then subtraction of two sets is defined as follows:

$$S - T = \{(p, \mu^P_{S-T}(p), \mu^I_{S-T}(p), \mu^N_{S-T}(p))|p \in U\}$$

where $\mu^P_{S-T}(p) = \vee[\mu^P_S(x) \wedge \mu^P_T(y)]$, $\mu^I_{S-T}(p) = \wedge[\mu^I_S(x) \vee \mu^I_T(y)]$, $\mu^N_{S-T}(p) = \wedge[\mu^N_S(x) \vee \mu^N_T(y)]$ and $p = x - y$.

Note 6: Let S and T be two PFS on the universal set U, where S and T are defined as follows:

$$S = \{(x, \mu_S^P(x), \mu_S^I(x), \mu_S^N(x))|x \in U\}$$

$$T = \{(y, \mu_T^P(y), \mu_T^I(y), \mu_T^N(y))|y \in U\}$$

Then the addition of two sets is defined as follows:

$$S * T = \{(p, \mu_{S*T}^P(p), \mu_{S*T}^I(p), \mu_{S*T}^N(p))|p \in U\}$$

where $\mu_{S*T}^P(p) = \vee[\mu_S^P(x) \wedge \mu_T^P(y)]$, $\mu_{S*T}^I(p) = \wedge[\mu_S^I(x) \vee \mu_T^I(y)]$, $\mu_{S*T}^N(p) = \wedge[\mu_S^N(x) \vee \mu_T^N(y)]$ and $p = x * y$.

Note 7: Let S and T be two PFS on the universal set U, where S and T are defined as follows:

$$S = \{(x, \mu_S^P(x), \mu_S^I(x), \mu_S^N(x))|x \in U\}$$

$$T = \{(y, \mu_T^P(y), \mu_T^I(y), \mu_T^N(y))|y \in U\}$$

Then the addition of two sets is defined as follows:

$$S/T = \{(p, \mu_{S/T}^P(p), \mu_{S/T}^I(p), \mu_{S/T}^N(p))|p \in U\}$$

where $\mu_{S/T}^P(p) = \vee[\mu_S^P(x) \wedge \mu_T^P(y)]$, $\mu_{S/T}^I(p) = \wedge[\mu_S^I(x) \vee \mu_T^I(y)]$, $\mu_{S/T}^N(p) = \wedge[\mu_S^N(x) \vee \mu_T^N(y)]$ and $p = x/y$.

Applications

In (Zhao et al. 2021) a new distance measure has been introduced based on picture fuzzy sets, which is more effective than the previous distance measures. It is useful to multi-attribute decision-making problems.

5.9 Rough Set

Consider an information system $P = (X, A)$, where X and A are non-empty, finite sets and X is called the universal set and A is a set of attributes. For every $a \in A$, the set B_a is the set of values (Pawlak 2002).

Indiscernible Relation:

Let S be a subset of A. A relation $I(S)$ on U is defined as follows:

$(x, y) \in I(S)$ if and only if $s(x) = s(y)$ for every $s \in S$,where $s(x)$ is the value of attribute s for element x. Now $I(S)$ is an equivalence relation. This relation is called indiscernible relation. The equivalence class containing x is denoted as $[x]$. $(X, [x])$ is called approximation space.

Upper and Lower Approximation Sets:

Let an information system $P = (X, A)$ be given. Upper and lower approximation sets for a subset U of X is defined as follows:

$$U^* = \{x \in X | [x] \cap U \neq \varnothing\}$$

$$U_* = \{x \in X | [x] \subseteq U\}$$

where $[x]$ is the equivalence class containing x.

Rough Set:

Let $P = (X, A)$ be an information system. For any target set $U \subseteq X$ and attribute subset S of A, U is called a rough set with respect to S if and only if $U^* \neq U_*$.

Operations:

Let U and V be two subsets of the universal set X.

Union:

$$(U \cup V)^* = U^* \cup V^*$$

$$(U \cup V)_* \supseteq U_* \cup V_*$$

Intersection:

$$(U \cap V)^* \subseteq U^* \cap V^*$$

$$(U \cap V)_* = U_* \cap V_*$$

Complement:

$$\overline{(U^*)} = (\overline{U})^*$$

$$\overline{(U_*)} = (\overline{U})_*$$

where $\overline{U} = X \backslash U$ (Zhang et al. 2016).

Applications

Rough set theory is very convenient in image processing and image segmentation. Image segmentation is the process of partitioning any image. Rough set theory is very useful in fast convergence and avoiding the local-minima problem. Image segmentation based on rough set theory is a stable and better process (Roy et al. 2014).

Soft set - Let U be the universal set and $P(U)$ be the power set of U. Let E be a set of parameters and $S \subseteq E$. A soft set (Molodtsov 1999) is pair (F, S), where $F : S \rightarrow P(U)$. Here F is called the soft set approximation of the soft set.

Applications

Soft set theory is very helpful for decision-making. In (Smithson and Oden 1999) authors described the decision-making algorithm. Here is an application of soft set theory to select a person for a job in a company. However, the soft theory is also useful in medical diagnosis (Yao 1998) and stock management (Pawlak 1985) etc.

Comparative Discussion

This section mainly discusses the similarities and differences between these types of FSs (Lee 2004). First, we notice that type-2 FS is an extension of FS. So FS can be considered as a subset of type-2 FS.

Now we consider IVFS and IFS. We can define an IVFS with the help of an IFS as follows:

Intuitionistic fuzzy set:

$$S = \{(p, \mu_S(p), \nu_S(p)) | p \in U\}$$

Interval-valued fuzzy set:

$$S = \left\{\left(p, \left[\mu_S^L(p), \mu_S^R(p)\right]\right) | p \in U\right\},$$

where $\mu_S^L(p) = \mu_S(p)$ and $\mu_S^R(p) = 1 - \nu_S(p)$.

The basic operations such as union, intersection, and complement also obey the above definition. So we can say that IVFS and IFS are equivalent.

Now we consider IFS and NS. An IFS can be considered a subset of an NS if we define it as follows:

IFS:

$$S = \{(p, \mu_S(p), \nu_S(p))|p \in U\}$$

Neutrosophic fuzzy set:

$$S = \{(p, T_S(p), F_S(p), I_S(p))|p \in U\},$$

where $T_S(p) = \mu_S(p)$, $F_S(p) = \nu_S(p)$ and $I_S(p) = 1 - (\mu_S(p) + \nu_S(p))$.

A PFS is different from NS since neutrality is not always the same as indeterminacy.

If we try to compare a BVFS $S = \{(p, \mu_S^P(p), \mu_S^N(p))|p \in U\}$ with an IFS with the condition $\mu_S^P(p) = \mu_S(p)$ and $\mu_S^N(p) = -\nu_S(p)$ Similar, but these are not since not-property is completely different from counter-property.

Again a BVFS is a special type of an m-polar valued fuzzy set. So BVFS can be considered as a subset of m-polar valued fuzzy set.

Here we are going to discuss the relation (Bustince et al. 2015) between the different types of FSs pictorially, as following Fig. 5.20.

5.10 Conclusion

This chapter surveyed the fact that why fuzzy set theory is required over classical set theory and how the world has benefited from fuzzy set theory over six decades. Here we have studied fuzzy sets and their extensions. The history of fuzzy set theory and its extensions are discussed elaborately. Set operations like union, intersection, and complement on these types of fuzzy sets have been illustrated pictorially. Their applications in our real-life problems have been provided. Few other applications of fuzzy sets theory can be found in Kalampakas et al. (2024); Mandal et al. 2024; Mandal et al. 2023; Mandal et al. 2022; Sahoo et al. 2022). Finally comparative relation of different types of fuzzy sets is provided. All mentioned extensions of the fuzzy set can be modeled to a single set satisfying all the conditions of the extensions. This can be assumed as the future scope of the study.

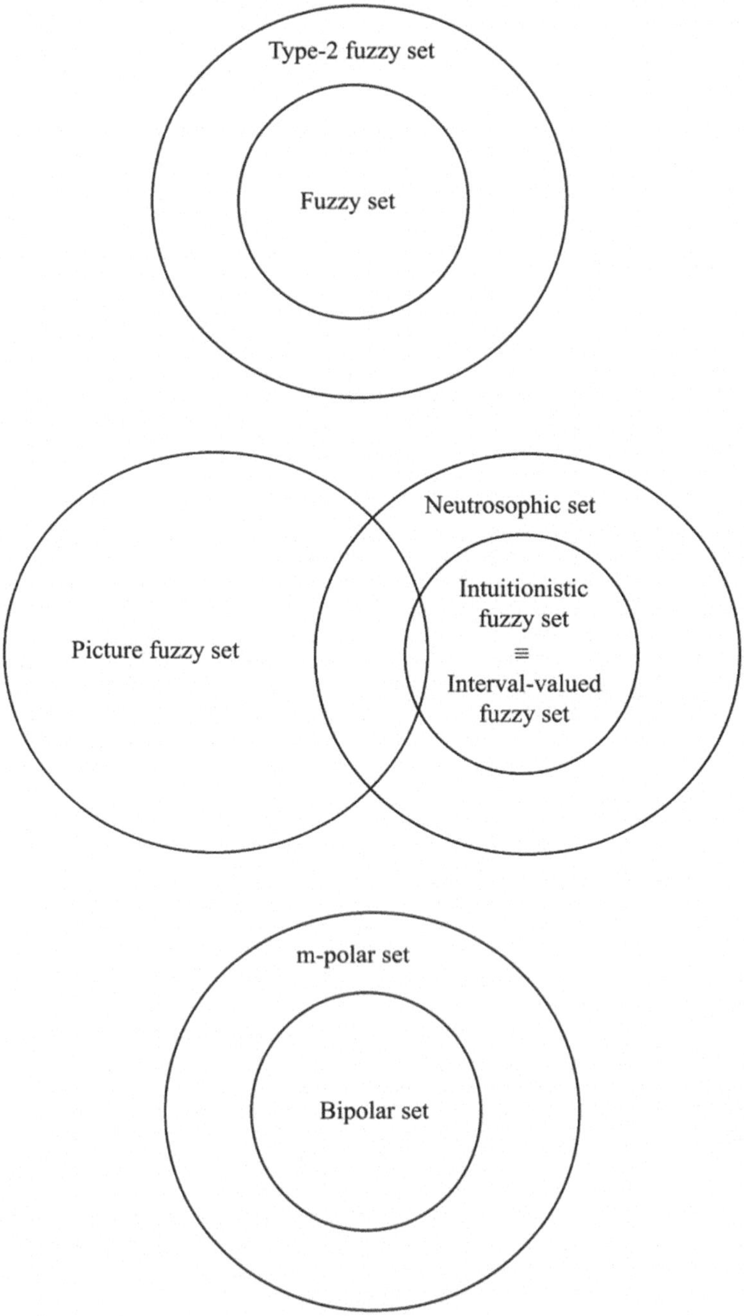

Fig. 5.20 Inclusion relationships between different types of FSs

Funding: No funds are available for this study.

Conflict of Interest: The authors declare that they have no conflict of interest to declare.

Ethical Approval: This article does not contain any studies with human participants or animals performed by any of the authors.

References

Alam J, Pandey MK (2014) Advance cyber security system using fuzzy logic. ACME J Manag IT 10(1):17–27
Atanassov KT (1986) Intuitionistic fuzzy sets. Fuzzy Sets Syst 20:87–96
Atanassov KT (1994) New operations defined over the intuitionistic fuzzy sets. Fuzzy Sets Syst 61(2):137–142
Bhattacharya A, Pal M (2021) Vertex covering problems of fuzzy graphs and their application in CCTV installation. Neural Comput Appl 33(11):5483–5506
Bhattacharya A, Pal M (2022) Fuzzy tree covering number for fuzzy graphs with its real-life application in electricity distribution system. Sādhanā 47(4):280
Bustince H, Barrenechea E, Pagola M, Fernandez J, Xu Z, Bedregal B, Montero J, Hagras H, Herrera F, De Baets B (2015) A historical account of types of fuzzy sets and their relationships. IEEE Trans Fuzzy Syst 24(1):179–194
Castillo O, Melin P, Kacprzyk J, Pedrycz W (2007) Type-2 fuzzy logic: theory and applications. In *2007 IEEE international conference on granular computing (GRC 2007)* (pp 145–145). IEEE
Chen J, Li S, Ma S, Wang X (2014) Polar fuzzy sets: an extension of bipolar fuzzy sets. Sci World J
Cuong BC, Kreinovich V (2013) Picture Fuzzy Sets-a new concept for computational intelligence problems. In *2013 third world congress on information and communication technologies (WICT 2013)* (pp 1–6). IEEE
Cuong BC, Kreinovich V (2014) Picture fuzzy sets. J Comp Sci Cybern 30(4):409–420
Deschrijver G, Cornelis C (2007) Representability in interval-valued fuzzy set theory. Internat J Uncertain Fuzz Knowl Based Syst 15(03):345–361
Dubois D, Prade H (1978) Operations on fuzzy numbers. Int J Syst Sci 9(6):613–626
Dutta P, Ganju S (2018) Some aspects of picture fuzzy set. Trans Razmadze Math Inst 172(2):164–175
Grattan-Guinness I (1976) Fuzzy membership mapped onto intervals and many-valued quantities. Math Log Q 22(1):149–160
Gulistan M, Yaqoob N, Elmoasry A, Alebraheem J (2021) Complex bipolar fuzzy sets: an application in a transport's company. J Intell Fuzzy Syst 40(3):3981–3997
Jahn KU (1975) Intervall-wertige Mengen. Math Nachr 68(1):115–132
Jana C, Pal M, Wang JQ (2020) Bipolar fuzzy Dombi prioritized aggregation operators in multiple attribute decision making. Soft Comput 24:3631–3646
Kahraman C, Öztayşi B, Çevik Onar S (2016) A comprehensive literature review of 50 years of fuzzy set theory. Intern J Comput Intell Syst 9(sup1):3–24
Kalampakas A, Samanta S, Bera J, Das KC (2024) A fuzzy logic inference model for the evaluation of the effect of extrinsic factors on the transmission of infectious diseases. Mathematics 12(5):648
Lee KM (2000) Bipolar-valued fuzzy sets and their operations. In *Proc. Int. Conf. on Intelligent Technologies, Bangkok, Thailand, 2000* (pp. 307–312)
Lee KM (2004) Comparison of interval-valued fuzzy sets, intuitionistic fuzzy sets and bipolar valued fuzzy sets. Fuzzy Logic Intell Syst 14:125–129

Mahapatra R, Samanta S, Pal M, Xin Q (2019) RSM index: a new way of link prediction in social networks. J Intel Fuzzy Syst 37(2):2137–2151
Mahapatra R, Samanta S, Pal M, Xin Q (2020) Link prediction in social networks by neutrosophic graph. Infinite Study
Mandal P, Samanta S, Pal M, Ranadive AS (2022) Three-way decision model under a large-scale group decision-making environment with detecting and managing non-cooperative behaviors in consensus reaching process. Artif Intell Rev 55(7):5517–5542
Mandal P, Samanta S, Pal M, Ranadive ASC (2023) Social network trust relationship environment based advanced ovarian cancer treatment decision-making model: an approach based on linguistic information with experts' multiple confidence levels. Expert Syst Appl 229:120407
Mandal P, Samanta S, Pal M (2024) Failure mode and effects analysis in consensus-based GDM for surface-guided deep inspiration breath-hold breast radiotherapy for breast cancer under the framework of linguistic Z-number. Inf Sci 658:120016
Mendel JM, John RB (2002) Type-2 fuzzy sets made simple. IEEE Trans Fuzzy Syst 10(2):117–127
Mizumoto M, Tanaka K (1976) Some properties of fuzzy sets of type 2. Inf Control 31(4):312–340
Mohammedali MN, Rasheed M, Shihab S, Rashid T, Hamed SHA (2021) Fuzzy set technique application: the solar cell. J Al-Qadisiyah Comp Sci Mathem 13(1):62
Molodtsov D (1999) Soft set theory—first results. Comput Math Appl 37(4–5):19–31
Pal M, Samanta S, Ghorai G (2020) Modern trends in fuzzy graph theory, pp. 7–93). Berlin: Springer
Pawlak Z (1985) Rough sets and fuzzy sets. Fuzzy Sets Syst 17(1):99–102
Pawlak Z (2002) Rough set theory and its applications. J Telecomm Inform Technol, 7-10
Pramanik T, Samanta S, Pal M (2016) Interval-valued fuzzy planar graphs. Int J Mach Learn Cybern 7:653–664
Roan NT, Smarandache F (2021) H-max single-valued neutrosophic distance measure in medical diagnosis. Intern J Neutros Sci (IJNS) 15(2)
Roy P, Goswami S, Chakraborty S, Azar AT, Dey N (2014) Image segmentation using rough set theory: a review. Intern J Rough Sets Data Analy (IJRSDA) 1(2):62–74
Sahoo L, Sen S, Tiwary K, Samanta S, Senapati T (2022) Modified floyd–warshall's algorithm for Maximum connectivity in wireless sensor networks under uncertainty. Discret Dyn Nat Soc 2022(1):5973433
Samanta S, Pal M (2015) Fuzzy planar graphs. IEEE Trans Fuzzy Syst 23(6):1936–1942
Samanta S, Pramanik T, Pal M (2016) Fuzzy colouring of fuzzy graphs. Afr Mat 27(1):37–50
Samanta S, Dubey VK, Sarkar B (2021) Measure of influences in social networks. Appl Soft Comput 99:106858
Sambuc R (1975) Fonctions ϕ-floues. Application l'aide au diagnostic en pathologie thyroidienne, in, Thesis Univ. Marseille, France
Smarandache F (1999) A unifying field in Logics: Neutrosophic Logic. In *Philosophy* (pp. 1–141). American Research Press
Smarandache F (2005) Neutrosophic set-a generalization of the intuitionistic fuzzy set. Intern J Pure Appl Mathem 24(3):287
Smithson M, Oden GC (1999) Fuzzy set theory and applications in psychology. Pract Appl Fuzzy Technol, 557–585
Szmidt E, Kacprzyk J (2001) Intuitionistic fuzzy sets in some medical applications. In *International conference on computational intelligence* (pp. 148–151). Berlin, Heidelberg: Springer Berlin Heidelberg
Wang H, Smarandache F, Zhang Y, Sunderraman R (2010) Single valued neutrosophic sets. Infinite study
Xu Z, Yager RR (2010) Intuitionistic fuzzy Bonferroni means. IEEE Trans Syst Man Cybern Part B (Cybernetics) 41(2):568–578
Yao YY (1998) A comparative study of fuzzy sets and rough sets. Inf Sci 109(1–4):227–242
Zadeh LA (1965) Information and control. Fuzzy Sets 8(3):338–353
Zadeh LA (1975) The concept of a linguistic variable and its application to approximate reasoning—I. Inf Sci 8(3):199–249

Zadeh LA (2008) Is there a need for fuzzy logic? Inf Sci 178(13):2751–2779
Zadeh LA (1996) Fuzzy sets. In *Fuzzy sets, fuzzy logic, and fuzzy systems: selected papers by Lotfi A Zadeh* (pp 394–432)
Zhang WR (1994) Bipolar fuzzy sets and relations: a computational framework for cognitive modeling and multiagent decision analysis. In *NAFIPS/IFIS/NASA'94. Proceedings of the First International Joint Conference of The North American Fuzzy Information Processing Society Biannual Conference. The Industrial Fuzzy Control and Intellige* (pp 305–309). IEEE
Zhang Q, Xie Q, Wang G (2016) A survey on rough set theory and its applications. CAAI Trans Intell Technol 1(4):323–333
Zhao R, Luo M, Li S (2021) A dynamic distance measure of picture fuzzy sets and its application. Symmetry 13(3):436

Chapter 6
Optimization in the New Age: Mathematical Pathways to Sustainable Futures

Savita Garg and Neetu Rani

Abstract This chapter elaborates on the integral role mathematics plays in shaping sustainable solutions for our evolving world. The mathematical significance and historical development of optimization and sustainability have been discussed. A classification of basic optimization problems has been provided. Emphasizing the symbiotic relationship between optimization and sustainability, it has been discussed how mathematical frameworks, from linear programming to stochastic modeling, provide tools for informed decision-making. Such models not only aim for optimal outcomes but also embed long-term viability. As global challenges amplify, including unpredictable environmental shifts, emerging methodologies, particularly at the intersection of machine learning and optimization, become crucial. Through exploring various mathematical concepts, their applications in sustainability, and future research directions, this article underscores the profound significance of mathematics in charting a sustainable path forward.

Keywords Sustainability · Optimization · Linear programming · Stochastic modeling · Multi-objective optimization

6.1 Introduction

Mathematics has always been a cornerstone in understanding and shaping our world. From early computations in agriculture to advanced algorithms today, it provides a language for innovation (Kline 1972). With growing global challenges like climate change and resource depletion, the importance of sustainability has

S. Garg
Department of Mathematics, Mukand Lal National College, Yamuna Nagar, Haryana 135001, India

N. Rani (✉)
Department of Mathematics, Shivaji College (University of Delhi), Delhi 110027, India
e-mail: anresearch2023@gmail.com

R. Arora and S. Arora (eds.), *Optimization Techniques for Sustainable Environment Under Uncertainty*, Engineering Optimization: Methods and Applications,
https://doi.org/10.1007/978-981-96-0389-3_6

become paramount (Rockström 2009). As a result, optimization—the art and science of making the best decisions given constraints (Winston 2004)—is progressively being harnessed to drive sustainable solutions.

Linear programming, for instance, is being utilized to optimize renewable energy production (Cai et al. 2009). Meanwhile, stochastic modeling helps in predicting and mitigating environmental uncertainties (Kall and Wallace 1994). As we stand at the brink of an era where decisions carry long-term repercussions, the amalgamation of sustainability with optimization becomes not just preferable, but imperative.

Mathematics, as the foundational backbone, is pivotal in this endeavor. Its evolving techniques offer nuanced ways to balance the scales between advancement and preservation (Nesterov 2018). First, the historical mathematical development of optimization and sustainability has been discussed.

Historical Development of Optimization and Sustainability: A Mathematical Journey:

Optimization and sustainability, both deeply rooted in history, have had their paths heavily influenced by mathematics.

Optimization:

The journey began with the ancient Greeks. Euclid's postulates laid the foundations of geometry (Boyer and Merzback 2011), with the notable equation for the distance between two points being

$$d = \sqrt{(p_2 - p_1)^2 + (q_2 - q_1)^2}$$

The seventeenth century witnessed the birth of calculus. Newton and Leibniz introduced the concept of derivatives, with the first derivative test, $g\prime(x) = 0$, becoming an essential tool for finding local maxima and minima of functions (Newton 1972).

Fast-forward to the twentieth century, the simplex method emerged for linear programming, where the goal was to maximize or minimize a linear objective function

$$l = a_1 p_1 + a_2 p_2 + \cdots + a_n p_n,$$

subject to linear constraints.

Nonlinear optimization algorithms, solving $g\prime(p) = 0$ where $g(p)$ is nonlinear, and integer programming, where solutions are restricted to integer values, expanded the boundaries of optimization (Dantzig 1963).

Sustainability:

With the pressing concern of exponential growth represented mathematically as $Q(t) = Q_0 e^{gt}$ (where $Q(t)$ is the population at time t, Q_0 is the initial population, g is the growth rate), and the need for sustainable models became evident.

The 1972 Limits to Growth report incorporated systems of differential equations to predict resources and populations. For example, a simple model might be

$$dR/dt = -cQ(t),$$

where R represents resources and c is a consumption constant (Meadows et al. 1972).

By the 1990s, the ecological footprint metric emerged. Given as

$$EF = \sum_{p=1}^{m} Area_p \times YF_p,$$

where $Area_p$ is the area for the p-th land type and YF_p its yield factor, it quantified human impact on the planet (Wackernagel and Rees 1998).

Intersection of Optimization and Sustainability:

Multi-objective optimization, which seeks solutions satisfying criteria like $g_1(p), g_2(p), \ldots, g_m(p)$, became crucial for balancing economic, environmental, and social goals in sustainability (Deb 2011).

In energy production, an optimization problem might be: maximize the efficiency.

$$E = useful\ output/total\ input,$$

of renewable energy systems while minimizing environmental impact EI.

Economic models in the Green Economy might be based on the principle of maximizing utility $U(c, l)$ where c represents consumption and l leisure, subject to sustainable resource constraints.

The "Mathematics of Tomorrow" encapsulates a future where sustainable solutions are optimized to benefit humanity and the planet. The significance of optimization and sustainability in mathematical language can be elaborated as given below.

Mathematical significance of optimization:

Optimization, in its purest sense, refers to the procedure of choosing the best possible solution from a set of feasible solutions (Bazaraa et al. 2013). Mathematically, this is often framed as:

$$minimize_{t \in D}\, g(t)$$

$$subject\ to\ p_i(t) \leq 0, i = 1, 2, 3, \ldots, m$$

$$q_j(t) = 0, j = 1, 2, 3, \ldots, n$$

where $g(t)$ is the objective function we wish to minimize (or maximize), D is the domain of the problem, often a subset of $\mathbb{R}^n$, $p_i(t)$ are the inequality constraints, $q_j(t)$ are the equality constraints, m is the number of inequality constraints, and n is the number of equality constraints.

Optimization lies at the heart of many mathematical problems. It can be seen in linear programming (Dantzig 1963), calculus (finding maxima and minima of functions), and even in machine learning algorithms where we aim to minimize a loss function (Goodfellow et al. 2016). Understanding optimization is crucial because real-world problems frequently involve making the best possible decision given a set of constraints, whether it be maximizing profits, minimizing costs, or optimizing resource allocation.

Sustainability in mathematical terms:

Sustainability, at its core, pertains to the enduring capacity of systems and processes (WCED 1987). In the context of mathematics, sustainability can be introduced as a constraint or set of constraints ensuring long-term viability.

Let us consider a simple hypothetical situation: Imagine a forest with a certain number of trees. We wish to harvest trees for timber, but if we deplete the forest too quickly, it will not regenerate at a sufficient rate, leading to eventual deforestation (FAO 2018). A sustainable optimization model for this scenario can be expressed as

$$Maximize_x P(x)$$

$$subject\ to\ r(x) \geq r^*$$

$$x \leq n$$

where $P(x)$ is the profit function depending on the number of trees harvested, $r(x)$ is the rate of forest regeneration after x trees have been harvested, r^* is a predefined minimum regeneration rate to ensure sustainability, n is the total number of trees available in the forest.

By introducing the regeneration rate constraint $r(x) \geq r^*$, it has been ensured that the forest is harvested in a manner that allows it to rejuvenate, thereby embracing the principle of sustainability.

In broader terms, integrating sustainability into mathematical models often involves introducing constraints that ensure resource preservation, minimal environmental impact, and long-term viability (Meadows et al. 1972). Optimization, when tailored to sustainability, can yield solutions that balance immediate needs with long-term survival and flourishing (Rockström et al. 2009b, a).

6.2 Basic Optimization Problems-Classifications

Optimization problems often vary in complexity and nature. However, there are some standard forms and basic classifications that help categorize them (Table 6.1).

Table 6.1 Classifications of optimization problems

Sr. No	Classification	Description
1	Linear Programming (LP)	It involves linear objective functions and linear constraints (Dantzig 1963)
2	Quadratic Programming (QP)	It is similar to linear programming but has a quadratic objective function (Nocedal and Wright 1999)
3	Nonlinear Programming (NLP)	In nonlinear programming, either the objective function, the constraints, or both are nonlinear (Bertsekas 1999)
4	Integer Programming (IP)	The variables in integer programming are restricted to integer values (Nemhauser and Wolsey 1988)
5	Mixed-Integer Programming (MIP)	A combination of LP and IP. Some variables are restricted to integer values while others can be continuous (Williams 2013)
6	Convex Optimization	A special subset of optimization where the objective is a convex function, and the feasible set is a convex set (Boyd and Vandenberghe 2004)
7	Combinatorial Optimization	It involves finding an optimal object from a finite set of objects. Examples include the Traveling Salesman Problem and Knapsack Problem (Papadimitriou and Steiglitz 1998)

6.3 Core Mathematical Concepts in Sustainable Optimization

- Linear Programming:

Mathematical formulation: Given a matrix A of size $m \times n$, vectors b of size m and c of size n, LP is defined as

$$Maximize\ z = c't$$

with constraints.

$$At \leq b\ and\ t \geq 0$$

where t is an n-dimensional decision vector. Here, prime represents the transpose operation.

To exemplify, let us develop a real-life problem based on the theme of sustainability, specifically focusing on sustainable energy management across a national grid with multiple energy sources.

Problem Statement: A nation has n different types of renewable energy sources. The objective is to determine the optimal energy drawn from each source, ensuring that the nation's energy demands are met while minimizing associated costs and emissions.

Mathematical formulation: Let e_i be the energy drawn from the *ith* source for $1 \leq i \leq n$.

Objective function (to be minimized):

$$z = \sum_{i=1}^{n} (c_i \times e_i + g_i \times f(e_i)),$$

where c_i is the cost of extracting energy from source i, g_i is the emission factor for source $i, f(e_i)$ is a function that describes the efficiency of energy extraction.

Constraints: For nation's demand D

$$\sum_{i=1}^{n} e_i = D.$$

For maximum and minimum extraction limits

$$L_i^{min} \leq e_i \leq L_i^{max} \; for \; 1 \leq i \leq n$$

To explain the solution method of a general case, assume $f(e_i)$ as quadratic (nonlinear) given by,

$$f(e_i) = \alpha_i e_i^2 + \beta e_i + \gamma.$$

Thus, the quadratic programming problem obtained is

$$Minimize \; z = \sum_{i=1}^{n} (c_i e_i + g_i(\alpha_i e_i^2 + \beta e_i + \gamma))$$

subject to the constraints,

$$\sum_{i=1}^{n} e_i = D \; and$$
$$L_i^{min} \leq e_i \leq L_i^{max} \; for \; 1 \leq i \leq n$$

Optimality condition: Putting the gradient of the Lagrangian equal to zero for each i gives

$$c_i + 2\alpha g_i e_i + \beta g_i - \lambda + \mu_i^{\min} - \mu_i^{\max} = 0,$$

where λ is the Lagrange multiplier, and.

$$\mu_i^{min}(\geq 0) \; and \; \mu_i^{max}(\geq 0)$$

are the multiplicities for the lower and upper bound constraints, respectively. To ensure that either the constraint is active or slack

$$\mu_i^{min}\left(e_i - L_i^{min}\right) = 0, \ and$$

$$\mu_i^{\max}(L_i^{\max} - e_i) = 0.$$

The obtained system of equations can be solved using numerical methods to find e_i's, λ, μ_i^{min} and μ_i^{max}. For this solution, one needs to check the original constraints, inequality constraints, and the values of e_i lying within the bounds L_i^{min} and L_i^{max}. If any of those checks fail, adjustments or iterative refinements may be needed.

- Stochastic Modeling:

Formulation of stochastic optimization problems: Stochastic optimization aims to find the best decision in the presence of uncertainty. If the expected value of a random variable r is $E[r]$, then for a random vector R, the optimization problem is expressed as

$$Minimize\ E\left[g(x, R)\right]$$

with constraints.

$$E[h_i(d, R)] \leq 0\ for\ 1 \leq i \leq m,$$

where g and h_i are functions defining the objective and constraints, respectively (Shapiro et al. 2014).

For real-life application, consider a problem related to supply chain optimization for a multinational corporation.

Problem statement: A multinational company operates in n different countries, producing various products using locally sourced materials. Each country has its demand, production costs, and tariffs. The company must decide how much to produce in each country, considering local demand, export potentials, and associated costs.

Mathematical formulation: Let P_i be the amount of product to be produced in country i (for $1 \leq i \leq n$), E_{ij} be the amount of product to be exported from country i to country j ($\forall i, j$ such that $i \neq j$ and $1 \leq i, j \leq n$), and R_i be the random variable representing the demand fluctuation in country i.

Objective function (Cost function to minimize):

$$C\left(\vec{P}, \vec{E}, \vec{R}\right) = \sum_{i-1}^{n}\left(c_iP_i + \sum_{j=1, j\neq i}^{n} t_{ij}E_{ij} - r_iE_{ii}\right)$$

subject to the constraints

$$P_i = \sum_{j=1}^{n} E_{ij} + E_{ii} = D_i(R_i)\ for\ 1 \le i \le n$$

$$E_{ij} \ge 0 \forall i, j$$

$$P_i \ge 0\ for\ 1 \le i \le n$$

where, c_i is the production cost in country i, t_{ij} is the tariff for exporting from country i to country j, r_i is the revenue per unit product sold in country i, E_{ii} is the amount of product i sold domestically (i.e., not exported), and $D_i(R_i)$ represents the demand in country i, which is a function of the random variable R_i. Here, the stochastic nature of the demand is captured through random variables R_i for each country.

Solution: To simplify, first consider the deterministic equivalents of the problem. Replace the random demand $D_i(R_i)$ with its expected value

$$\overline{D_l} = E[D_i(R_i)].$$

For each country i,

$$P_i - \left(\sum_{j=1}^{n} E_{ij} + E_{ii} \right) = \overline{D_i}$$

which gives the production and export relationships.

Using the deterministic equivalents, the objective is

$$Minimize \sum_{i-1}^{n} \left(c_i P_i + \sum_{j=1, j \ne i}^{n} t_{ij} E_{ij} - r_i E_{ii} \right)$$

For optimal solutions, the equations are

$$\frac{\partial C}{\partial P_i} = c_i - \lambda_i = 0$$

$$\frac{\partial C}{\partial E_{ij}} = t_{ij} - \mu_i = 0$$

where λ_i is the Lagrange multiplier associated with the constraint for country i, μ_i is another Lagrange multiplier for the export constraint from country i. The solution of the above equations along with the constraints provided gives the solution that must satisfy the constraints,

$$P_i - \left(\sum_{j=1}^{n} E_{ij} + E_{ii} \right) \geq 0 \forall i$$

and

$$E_{ij} \geq 0 \,\forall\, i,\, j$$

If any constraint is not met, adjustments to the solutions will be required. Once a deterministic solution is obtained, one can adjust for stochasticity by considering various scenarios, i.e., one might evaluate the performance of this solution under different realizations of the random variables R_i.

- Multi-objective Optimization:

Weighted sum approach: This approach is used to combine multiple objectives, $g_i(1 \leq i \leq m)$ into a single objective $g(t) = \sum_{i=1}^{m} w_i g_i(t)$, where w_i are user-defined weights for each objective (Deb 2011).

For real-life application, consider a generalized problem related to urban planning.

Generalized problem statement: Urban planners must consider n different objectives while planning the land use of a city. Each of these objectives relates to a real-life consideration, such as environmental impact, economic gain, maintenance costs, societal well-being, and more.

Mathematical formulation:

$$For\ 1 \leq i \leq n,$$

$$g_i(t) = f_i(\overrightarrow{v})$$

where f_i represents the *ith* objective function, $\overrightarrow{v} = [v_1, v_2, \ldots, v_m]$ is a vector of decision variables, representing different facets of the urban plan, like commercial area, green space, residential area, infrastructure area, etc.

Combined objective is

$$g(t) = \sum_{i=1}^{n} w_i \times f_i(\overrightarrow{v})$$

where w_i are user-defined weights for each objective function. The problem may have a set of p constraints given by

$$h_k\left(\overrightarrow{v}\right) = 0\ for\ 1 \leq k \leq p$$

Applying condition for optimal solution for each $1 \leq j \leq m$ gives

$$\frac{\partial g(t)}{\partial v_j} = \sum_{i=1}^{n} w_i \frac{\partial f_i}{\partial v_j} = 0$$

This gives a system of m equations. To better understand the solution procedure, let us frame a more specific problem with 5 objectives, each representing different facets of urban planning.

Problem statement: Urban planners are tasked with optimizing land use for a new city district. The land is to be divided among residential areas v_1, commercial areas v_2, green spaces v_3, public infrastructure v_4, and recreational facilities v_5. The goal is to optimize environmental health, economic viability, resident well-being, infrastructure efficiency, and recreational availability.

Objectives of the problem are,

Environmental impact (Minimize):

$$g_1(t) = a_1 v_1 + a_2 v_2 - a_3 v_3$$

where coefficients represent the environmental burden (or benefit) of each land type.

Economic gain (Maximize):

$$g_2(t) = b_1 v_2 + b_2 v_4 - b_3 v_5$$

where the economic benefit of commercial areas and public infrastructure, minus the upkeep of recreational facilities, is taken into account.

Resident well-being (Maximize):

$$g_3(t) = c_1 v_1 + c_2 v_3 + c_3 v_5$$

promoting residential areas, green spaces, and recreational facilities.

Infrastructure efficiency (Maximize):

$$g_4(t) = d_1 v_4 - d_2 v_1 v_2$$

accounting for the efficiency of public infrastructures but also considering the complexity induced by interfacing commercial and residential areas.

Recreational availability (Maximize):

$$g_5(t) = e_1 v_5 - e_2 v_2$$

promoting recreational areas while considering the space taken by commercial areas.

Combined objective:

$$g(t) = \sum_{i=1}^{5} w_i \times g_i(t)$$

i.e.,

$$\begin{aligned} g(t) &= w_1(a_1v_1 + a_2v_2 - a_3v_3) + w_2(b_1v_2 + b_2v_4 - b_3v_5) + w_3(c_1v_1 + c_2v_3 + c_3v_5) \\ &\quad + w_4(d_1v_4 - d_2v_1v_2) + w_5(e_1v_5 - e_2v_2) \end{aligned}$$

Consider the total land area as L, i.e.,

$$v_1 + v_2 + v_3 + v_4 + v_5 = L$$

And assume the following specific requirements for the city given as
Residential area should be at least 30%:

$$v_1 \geq 0.3L$$

Green spaces should be at least 20%:

$$v_3 \geq 0.2L$$

Recreational facilities should not exceed 10%:

$$v_5 \leq 0.1L$$

Lagrangian for the problem will be

$$\begin{aligned} \mathcal{L} &= w_1(a_1v_1 + a_2v_2 - a_3v_3) + w_2(b_1v_2 + b_2v_4 - b_3v_5) + w_3(c_1v_1 + c_2v_3 + c_3v_5) \\ &\quad + w_4(d_1v_4 - d_2v_1v_2) + w_5(e_1v_5 - e_2v_2) + \lambda(L - v_1 - v_2 - v_3 - v_4 - v_5) \end{aligned}$$

Optimality condition gives

$$\frac{\partial \mathcal{L}}{\partial v_1} = w_1a_1 - w_4d_2v_2 + w_3c_1 - \lambda = 0$$

$$\frac{\partial \mathcal{L}}{\partial v_2} = w_1a_2 + w_2b_1 - w_4d_2v_1 - w_5e_2 - \lambda = 0$$

$$\frac{\partial \mathcal{L}}{\partial v_3} = -w_1a_3 + w_3c_2 - \lambda = 0$$

$$\frac{\partial \mathcal{L}}{\partial v_4} = w_2b_2 - \lambda = 0$$

$$\frac{\partial \mathcal{L}}{\partial v_5} = -w_2b_3 + w_3c_3 + w_5e_1 - \lambda = 0$$

Solution of these equations along with the constrain on total land gives the optimal solution as $v_1^*, v_2^*, v_3^*, v_4^*$, and v_5^*. The next step is the verification of inequality constraints described as if $v_1^* < 0.3L$, then the solution violates the residential areas constraint, so increase v_1^* to$0.3L$. Since the total land area is fixed, the adjustment of the other variables would be required, possibly reducing one or more of the others to maintain the total land constraint while maximizing the objective function. Similar verification and adjustment if needed is to be done for the green space and recreational area constraints. It is possible, especially with complex multi-objective optimization problems, that after adjustment, the new solution might not be optimal. Instead, it would represent a feasible solution that respects constraints while being as close to optimal as possible.

6.4 Challenges and Limitations

Ill-posed problems and non-uniqueness: In optimization, a problem is termed as ill-posed if it does not satisfy one of the three Hadamard criteria—existence, uniqueness, and stability. Consider an optimization problem with objective $g(t)$ and constraints $h(t)$. If there does not exist a solution t^* such that $h(t^*) = 0$, the problem violates the existence criterion. If multiple distinct solutions $t_1^*, t_2^*, \ldots$ all satisfy $h(t) = 0$ and have the same optimal value for $g(t)$, the problem lacks uniqueness. To exemplify, consider an urban planning problem given as, problem context: Urban planners in a rapidly growing city want to design a new bus route for a burgeoning business district.

Hypothesis: Worker density in the business district varies throughout the day as

$$D(t) = Pe^{\frac{-t^2}{16}}\cos(t) + Q$$

where t is the time of day (in hours, $0 \leq t \leq 24$), and P, Q are constants.

Mathematical formulation: Define bus dispatch frequency as $f(t)$, where waiting time $W(t)$ is given by,

$$W(t) = K\frac{D(t)}{f(t)}$$

where K is a constant. The objective function is given as,

$$Minimize\ J = \int_6^{20} W(t)dt$$

This integral represents the cumulative waiting time throughout the business day. The objective function is subjected to the constraints,

$$f(t) \leq f_{max}$$

$$Operational\ cost: \int_{6}^{20} Cf(t)dt \leq B$$

where C is the cost per bus per hour and B is the budget.

Existence of solution: The expression for $D(t)$ guarantees that there are peak times when worker density is high. An optimal frequency $f(t)$ exists that minimizes the waiting time.

Uniqueness of solution: The function $f(t)$ is required that minimizes J. With constraints, it is likely there will be multiple solutions that achieve near-minimum values. The Lagrangian for this optimization problem is set as

$$\mathcal{L}(f(t), \lambda) = \int_{6}^{20} K\frac{D(t)}{f(t)}dt + \lambda\left(\int_{6}^{20} Cf(t)dt - B\right)$$

where λ is the Lagrange multiplier associated with the budget constraint. Taking the first variation with respect to $f(t)$ and setting it to zero provides,

$$\frac{\partial \mathcal{L}}{\partial f(t)} = -K\frac{D(t)}{f(t)^2} + \lambda C = 0$$

which is a functional differential equation that needs to be solved subject to the given constraints. Solving this could yield multiple functions $f(t)$ that satisfy the criteria, indicating a non-uniqueness in the solution.

Sensitivity analysis in the face of data uncertainty: Given an optimization problem with objective function $g(d, p)$, where d is the decision variable and p is a parameter. Sensitivity analysis examines how changes in p affect the optimal solution d^*. The sensitivity, $S(p)$, can be defined as $S(p) = \partial d^* / \partial p$. For sustainability optimization, uncertain data might come from varied sources, like unpredictable weather for renewable energy. If $S(p)$ is large, even slight changes in p can lead to significant shifts in d^*, indicating the sensitivity of the solution (Saltelli et al. 2008). To better understand the concept, consider a renewable energy production problem given as,

Problem context: Consider a wind turbine in a location where wind speed varies unpredictably. The power generated by the turbine is given by the function,

$$P(d, v) = dv^3 e^{-\frac{d^2}{v}}$$

where P is the power generated, d is the design parameter related to blade length, v is the wind speed.

Objective: Maximize $P(d, v)$ with respect to d for a given v.

Optimality condition gives,

$$\frac{\partial P}{\partial d} = v^3 e^{\frac{-d^2}{v}} - 2d^2 v^2 e^{\frac{-d^2}{v}} = 0$$

which gives d^* for any given v, i.e., $d^*(v)$. The Sensitivity is evaluated as

$$S(v) = \frac{\partial d^*}{\partial v}$$

Obtaining an analytical expression for $S(v)$ directly might be complex. However, numerical methods can be used to determine how d^* changes with slight variations in v. If $|S(v)|$ is large, then the system is highly sensitive to changes in wind speed v.

6.5 Future Research Directions

Intersection of machine learning and optimization for sustainability: Modern sustainability challenges often combine optimization with machine learning. Consider a predictive model M trained using data D to predict sustainable outcomes. The optimization task then becomes- Minimize $L(d, M)$, where L is a loss function measuring discrepancies between model predictions and desired sustainable outcomes, and d represents decisions to optimize.

Probabilistic models for predicting sustainable outcomes: Given random variables $r_1, r_2, \ldots, r_n$ representing environmental or social factors, a probabilistic model P estimates the likelihood of a sustainable outcome s given as-$P(s|r_1, r_2, \ldots, r_n)$. This formulates sustainability as a conditional probability problem, aiding decision-making under uncertainty.

Robust optimization: This aims to find solutions that perform well under the worst-case scenarios of uncertainty (Ben-Tal and Nemirovski 1998). For a decision variable d and uncertain parameter p, the robust model becomes Minimize$_{\max p} f(d, p)$.

Chance-constrained optimization: This optimization ensures that constraints are satisfied with a certain probability β and is given as $P(h(d, p) \leq 0) \geq \beta$. This captures the probabilistic nature of constraints under uncertainty.

6.6 Conclusion

Mathematics, with its rich tapestry of techniques and methodologies, has emerged as a cornerstone in our collective pursuit of a sustainable future. The article vividly underscores the powerful blend of optimization, a branch of mathematics, and the overarching goal of sustainability. As we strive to address global challenges, this union enables us to make informed decisions that serve both immediate and long-term benefits. Using optimization, we can deftly navigate challenges, ensuring that our actions today don't compromise the resources of tomorrow. Techniques like linear programming and stochastic modeling, traditionally reserved for theoretical or commercial applications, now find profound resonance in tackling real-world sustainability issues. As we stand on the cusp of monumental global changes, mathematics acts as our compass, pointing us toward choices that harmonize our needs with the planet's well-being. This is not just about using resources wisely; it's about reshaping our approach to ensure lasting prosperity. In essence, the interplay of mathematical sustainability and optimization illuminates our path, assuring a world where progress aligns seamlessly with preservation.

References

Bazaraa MS, Sherali HD, Shetty CM (2013) Nonlinear programming: theory and algorithms. John Wiley & Sons

Ben-Tal A, Nemirovski A (1998) Robust convex optimization. Math Oper Res 23(4):769–805. https://doi.org/10.1287/moor.23.4.769

Bertsekas DP (1999) Nonlinear programming. Athena scientific

Boyd S, Vandenberghe L (2004) Convex optimization. Cambridge University Press

Boyer CB, Merzbach UC (2011) A history of mathematics. John Wiley & Sons

Cai YP, Huang GH, Lin QG (2009) An optimization-model-based interactive decision support system for regional energy management systems planning under uncertainty. Energy 34(9):1304–1314. https://doi.org/10.1016/j.eswa.2008.02.036

Dantzig GB (1963) Linear programming and extensions. Princeton University Press. https://doi.org/10.1515/9781400884179

Deb K (2011) Multi-objective optimization using evolutionary algorithms. John Wiley & Sons. https://doi.org/10.1007/978-0-85729-652-8_1

FAO (2018) The state of the world's forests 2018. Food and Agriculture Organization of the United Nations

Goodfellow I, Bengio Y, Courville A (2016) Deep learning. MIT Press

Kall P, Wallace SW (1994) Stochastic programming. John Wiley & Sons

Kline M (1972) Mathematical thought from ancient to modern times. Oxford University Press

Meadows DH, Meadows DL, Randers J, Behrens III WW (1972) The limits to growth. New York, 102

Nemhauser GL, Wolsey LA (1988) Integer and combinatorial optimization. John Wiley & Sons

Nesterov Y (2018) Lectures on convex optimization. Springer. https://doi.org/10.1007/978-3-319-91578-4

Newton I (1972) Philosophiæ Naturalis principia mathematica. Joseph Streater, London

Nocedal J, Wright S (1999) Numerical optimization. Springer Sci Bus Media. https://doi.org/10.1007/0-387-22742-3_18

Papadimitriou CH, Steiglitz K (1998) Combinatorial optimization: algorithms and complexity. Courier Corporation

Rockström J et al (2009a) A safe operating space for humanity. Nature 461(7263):472–475. https://doi.org/10.1038/461472a

Rockström J, Steffen W, Noone K, Persson Å, Chapin FS, III, Lambin E, Lenton TM, Scheffer M, Folke C, Schellnhuber H, Nykvist B, De Wit CA, Hughes T, van der Leeuw S, Rodhe H, Sörlin S, Snyder PK, Costanza R, Svedin U, Falkenmark M, Karlberg L, Corell RW, Fabry VJ, Hansen J, Walker B, Liverman D, Richardson K, Crutzen P, Foley J (2009) Planetary boundaries: exploring the safe operating space for humanity. Ecol Soc 14(2):32. https://www.jstor.org/stable/26268316

Saltelli A, Ratto M, Andres T, Campolongo F, Cariboni J, Gatelli D, Saisana M, Tarantola S (2008) Global sensitivity analysis: the primer. John Wiley & Sons

Shapiro A, Dentcheva D, Ruszczyński A (2014) Lectures on stochastic programming: modeling and theory. SIAM. https://doi.org/10.1137/1.9781611973433.ch5

Wackernagel M, Rees W (1998) Our ecological footprint: Reducing human impact on the Earth. New Society Publishers

WCED (World Commission on Environment and Development) (1987) Our common future (The Brundtland Report). Oxford University Press

Williams HP (2013) Model building in mathematical programming. John Wiley & Sons

Winston WL (2004) Operations research: applications and algorithms. Brooks/Cole-Thomson Learning

Chapter 7
Multi-objective Routing Optimization of the Traveling Salesman Problem Using Feasibility-Enhanced Particle Swarm Optimization Algorithm

Salar Farahmand-Tabar and Parastoo Afrasyabi

Abstract This research focuses on the multi-objective optimization of the Traveling Salesman Problem (TSP) considering the cities of Iran. The objective is to minimize both the length of the path and the traffic congestion. To achieve this, a Feasibility-Enhanced Particle Swarm Optimization (FEPSO) algorithm is implemented. The FEPSO algorithm incorporates a feasibility-based approach to enhance the performance of the traditional Particle Swarm Optimization (PSO) algorithm in handling multiple objectives. The algorithm is designed to find optimal routes for salespeople visiting all cities in Iran while considering the shortest path, traffic congestion, safety, and facility. The results demonstrate the effectiveness of the FEPSO algorithm in obtaining well-balanced solutions in terms of path length and traffic congestion for the TSP.

Keywords Multi-objective optimization · Traveling salesman problem · Feasibility-enhanced particle swarm optimization · Routing problem

7.1 Introduction

Routing is one of the classic problems that was formally addressed for the first time by Mr. Hamilton, a famous mathematician, under the title of Traveling Salesman (TSP) (Gutin and Yeo 2001). The traveling salesman problem is one of the famous problems in optimization, which has many applications in various sciences. For example, it is used to find the optimal path for moving the drill-to-drill holes in printed circuit boards

S. Farahmand-Tabar (✉)
Faculty of Engineering, Department of Civil Engineering, University of Zanjan, Zanjan, Iran
e-mail: farahmandsalar@znu.ac.ir

P. Afrasyabi
Faculty of Geodesy and Geomatics Engineering K.N, Toosi University of Technology, Tehran, Iran

R. Arora and S. Arora (eds.), *Optimization Techniques for Sustainable Environment Under Uncertainty*, Engineering Optimization: Methods and Applications,
https://doi.org/10.1007/978-981-96-0389-3_7

(Duman and Or 2004). TSP is a complex optimization problem of NP-hard order (Jourdan et al. 2009), which has a wide solution space. NP-hard is a set of all decision problems that can be verified by a non-deterministic Turing machine using a non-deterministic algorithm, in the worst-case polynomial time (Ke-Lin 2016). Unlike deterministic methods such as Conjugate Gradient Method (Farahmand-Tabar et al. 2023a), non-deterministic methods such as meta-heuristic algorithms are used to solve the traveling salesman problem. Among the well-known meta-heuristic algorithms used to solve the traveling salesman problem are genetic algorithms (Ashtari et al. 2021; Farahmand-Tabar and Ashtari 2020, 2023; Farahmand-Tabar 2023a) ant colony (Farahmand-Tabar and Shirgir 2023a, b; Farahmand-Tabar 2023b), and particle swarm optimization (Fang et al. 2007; Kaveh et al. 2020a; Kaveh and Mesgari 2019a; Khishe et al. 2017; Mosavi et al. 2016a).

Meta-heuristic algorithms are classified into two categories based on single answer and based on population (Lotfy et al. 2020; Mosavi et al. 2016b; Mousavi et al. 2016; Kaveh and Mesgari 2019b). The multi-objective particle swarm optimization (MOPSO) algorithm is based on population intelligence. Usually, the MOPSO algorithm is presented for continuous problems; but for discrete problems, it works very strongly by applying changes and generalization to the discrete state. The MOPSO algorithm searches the problem space and finds optimal solutions (Shi and Eberhart 1998). In the MOPSO, the Pareto chart and the Pareto solution front are used to rank the solutions. The non-dominant solutions are determined by the Pareto chart. Graph and adjacency matrix are used to model the TSP problem. A particle represents a set of edge sequences around the Hamiltonian. In this problem, the permutation of the edges is important and the repeated edge should not be included in the answer (Applegate et al. 2006).

In many routing problems, the user pursues several goals instead of one goal, and some of these goals may be inconsistent; the misalignment between the objectives makes it impossible to transform the objectives linearly into a more general objective function. An example of such misalignment goals is finding the shortest route and at the same time the least expensive and shortest route time, or a route with the lowest cost and least traffic. Most of today's systems focus on one goal, which will cause them to ignore some other aspects of routing. Paying attention only to the shortest route will lead to not paying attention to the traffic on the route, and perhaps the shortest route is the high-traffic route. Often, the minimization of more than one objective function is considered, or in other words, multi-objective decision-making problems arise here (Jozefowiez et al. 2008). TSP is one of these issues that should be considered in a multi-objective manner. The TSP problem involves the shortest path that passes through each city once and only once. That is, in the basic traveling salesman problem, the goal is to minimize the path distance. But in some cases, other goals become important for the user; Goals that are inconsistent with each other. In such complex cases of optimization, using and improved metaheuristics is required (Farahmand-Tabar and Shirgir 2023c, d, e, f; Farahmand-Tabar and Babaei 2023; Shirgir et al. 2023; Farahmand-Tabar and Sadrekarimi 2023; Farahmand-Tabar and Rashid 2023; Farahmand-Tabar et al. 2023b, 2024; Farahmand-Tabar 2023c).

In this study, for the traveling salesman problem, multiple objectives such as route length and traffic, safety, and facility are considered. Also, the multi-objective traveling salesman problem is intrinsically discrete. So, it is better to use discrete algorithms to solve it (Clerc 2004). Therefore, among different methods (Farahmand-Tabar and Shirgir 2024; Farahmand-Tabar 2024; Farahmand-Tabar and Afrasyabi 2024) to solve the discrete TSP problem, the development of the multi-objective feasibility-enhanced particle swarm optimization (MOFEPSO) algorithm has been used. Finally, the obtained results are compared with the MOPSO and non-dominated sorting genetic algorithm (NSGA-II).

7.2 Background Studies on Multi-objective Traveling Salesman Problem

In general, a lot of research has been done to solve the problem of the shortest path or to find the optimal path, and there are many methods to obtain the shortest path or the optimal path. These methods are divided into two general categories: deterministic methods and random methods. In deterministic methods, there is a definite and final solution to the problem, but in stochastic methods, we have a set of solutions. In the TSP problem, the goal is to find a path between two source and destination nodes of a network so that the total weight of the edges of the path is minimized (Bellman 1958). PSO is a meta-heuristic algorithm that is simple yet powerful in solving problems (Kaveh et al. 2019, 2020). Kennedy and Eberhart presented the first discrete version of PSO in binary form (Kennedy and Eberhart 1997). Clerc gave a summary of the PSO method for solving TSP problems (Clerc 2004). Hendtlass implemented the PSO algorithm to solve small-sized traveling salesman problems and improved the performance of the PSO algorithm (Hendtlass 2003). Wang et al. improved the redefinition of particle swarm optimization operators by introducing the concepts of "swapping operator" and "swapping sequence" of the PSO algorithm and solved the traveling salesman problem by the improved algorithm (Wang et al. 2003).

Stochastic methods include meta-heuristic algorithms. Among meta-heuristic algorithms, we can mention ant algorithm, genetics, NSGA-II, artificial bee colony, and particle mass optimization. The most famous and basic meta-heuristic algorithm is genetic algorithm. The NSGA-II algorithm is an upgrade of the single-objective genetic algorithm to the multi-objective state. NSGA-II algorithm is one of the most powerful meta-heuristic algorithms in solving multi-objective problems. Pires and his colleagues have used the elite non-dominant sorting genetic algorithm (NSGA-II) with local search to find computational solutions to a multi-objective reactive power compensation model. This performance is improved by using a local search scheme in NSGA-II, which is specifically tuned to the problem characteristics. Using local search in the NSGA-II operational framework in this problem improves convergence and ensures that the obtained solutions are well distributed (Pires et al. 2012).

Bolaños et al. have solved a multi-objective version of the multiple traveling salesman problem (MOmTSP). In particular, two objectives are considered: minimizing the total distance traveled and balancing the working hours of the traveling salesmen. This problem is formulated as an integer multi-objective optimization model. NSGA-II algorithm has been improved to solve MOmTSP. The solution design allows a set of ordered solutions to be found in Pareto fronts, taking into account the concept of dominance. Experiments on real-world examples and cases adapted from the literature demonstrate the effectiveness of the algorithm proposed by Bolaños and colleagues (Bolaños et al. 2015).

In their paper, Shuai et al. presented an effective multi-objective evolutionary algorithm for solving multiple traveling salesman problem. In order to obtain the minimum visited distance and the minimum range between all vendors, some new display operators have designed crossovers and jumps to improve local and global search behaviors. Then, they have used the NSGA-II framework to find the appropriate convergence and diversity of non-dominated solutions. The proposed algorithm has been compared with several advanced approaches and the comparison results show that the proposed algorithm is effective and efficient for solving multiple traveling salesman problems (Shuai et al. 2019).

7.3 Optimization Problem

Solving multi-objective optimization problems using deterministic methods is very time-consuming. Stochastic methods can find the solution close to the optimal solution in a suitable time interval. In this section, the concepts related to the multi-objective traveling salesman problem, the NSGS-II algorithm, and the multi-objective feasibility-enhanced particle swarm optimization (MOFEPSO) algorithm (Hasanoglu and Dolen 2018) will be explained.

7.3.1 Multi-objective Traveling Salesman Problem

The traveling salesman problem is perhaps the most famous optimization benchmark problem. In this problem, there is a set of cities, all cities have roads to each other. In this problem, we are looking for a route that starts from a city and passes through all the cities once and only once (Hamilton cycle) and finally returns to the first city. Also, the distance traveled should be the minimum possible. The traveling salesman problem is used to solve many real-world problems. But the basic problem of the single-purpose traveling salesman (only the length of the route should be minimized) is not the solution to many problems, and in many problems, more than one goal is considered. Here, in addition to the route length, the traffic function is also assumed as an objective. This means that it is an optimal route that has less distance and less traffic (Coello et al. 2007).

To define the traveling salesman problem, a graph is used where the number of nodes is equal to the number of n cities and the number of edges is equal to $n(n-1)/2$. Also, the proximity matrix is defined with size nxn, which represents the connection between cities. The multi-objective traveling salesman problem in this study has two objective functions: distance and traffic; to define the objective functions, two matrices equal to the size of the adjacency matrix are defined, which represent the weight of each edge. In the distance matrix, the distance between two cities i and j is equal to each area of the matrix. In the multi-objective traveling salesman problem, the size of the search space is equal to $n!$. In this research, the optimal solution in the itinerant salesman problem is subject to the two goals of distance traveled and route traffic. The optimal answer is a route that has less length and less traffic. A longer route may have less traffic or a shorter route may have more traffic. The length and traffic volume of the route are equal to the total length and traffic of each of the edges of the route. In Eq. (7.1), the optimality function is expressed in mathematical language.

$$Fitness\ Function = Min \sum_{i,j=1}^{n} D_{i,j} + T_{i,j} \tag{7.1}$$

where i and j are the number of the city $D_{i,j}$, the length of the edge between two cities i and j and the edge traffic between two cities i and j is equal to $T_{i,j}$.

7.4 Feasibility-Enhanced Particle Swarm Optimization

Multi-objective constrained optimization issues can be described in a general manner as seeking to minimize multiple objectives simultaneously while adhering to specific constraints:

$$y = \mathbf{f}(\boldsymbol{x}) \equiv [f_1(\boldsymbol{x}) f_2(\boldsymbol{x}) \cdots f_K(\boldsymbol{x})]^T$$

$$g_m(x) \leq 0, m \in \mathbb{N}_{>0}^{\leq M} \tag{7.2}$$

$$h_p(x) = 0, p \in \mathbb{N}_{>0}^{\leq P}$$

Here, K represents the number of objective functions; the decision vector $x = [x_1 x_2 \cdots x_N]^T \in \mathbb{R}^{N\times 1}$ contains N variables. M and P denote the counts of inequality and equality constraints, respectively. Additionally, each decision variable is subject to lower and upper limits, ensuring that:

$$x_n^L \leq x_n \leq x_n^U, n \in \mathbb{N}_{>0}^{\leq N} \tag{7.3}$$

Typically, equality constraints are converted into inequality constraints using the formulation $|h_p(x)| - \varepsilon \leq 0$. It's important to note that MOFEPSO is designed to handle only inequality constraints. For a visual representation of MOFEPSO's operation, refer to Fig. 7.1. The subsequent sections will provide further insights into the methodology outlined in Fig. 7.1.

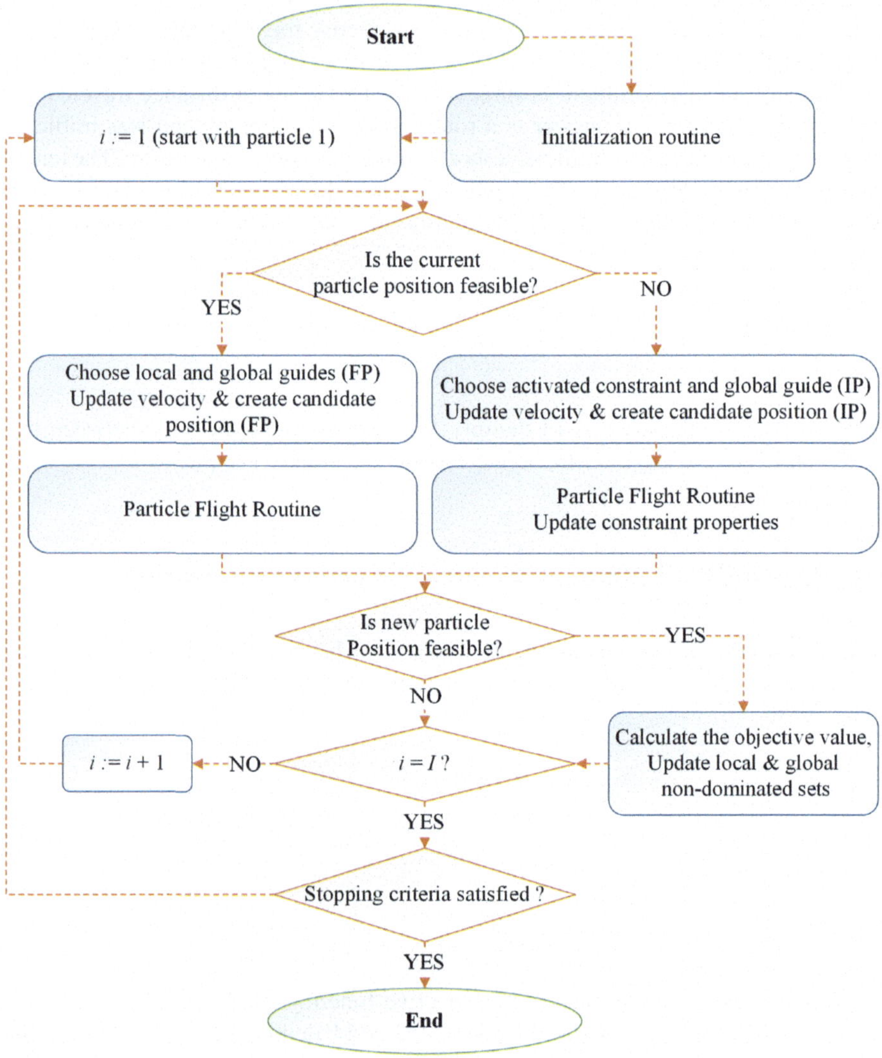

Fig. 7.1 Flowchart of MOFEPSO (Hasanoglu and Dolen 2018)

7.4.1 Initialization

The tasks preceding the main MOFEPSO iterations, labeled as the 'Initialization procedure in Fig. 7.1, involve several operations. The steps of this initialization routine are outlined in Fig. 7.2. The initial phase of MOFEPSO involves randomly initializing the position matrix X and the velocity matrix V. These matrices can be represented as follows:

$$\begin{aligned} X &= [x_{i,n}] \in \mathbb{R}^{1\times N} \\ V &= [v_{i,n}] \in \mathbb{R}^{1\times N} \end{aligned} \tag{7.4}$$

Here, $i \in \{1, 2, ..., I\}$ and$n \in \{1, 2, ..., N\}$. I represent the number of particles in the swarm, while N is the count of decision variables. Consequently, each row of X in the matrix represents the position vector of an individual particle in the swarm.

$$X_{i,*} = [x_{i,1}x_{i,2} \dots x_{i,N}] \in \mathbb{R}^{1\times N} \tag{7.5}$$

Each row in the matrix represents the transposed position vector of the corresponding particle. Various methods can be employed for random position initialization. For this article, the Latin Hypercube Sampling Method is utilized, leveraging its simultaneous stratification property on all input dimensions. As a result, this technique ensures a well-distributed initial population across all dimensions.

The initialization of the velocity matrix is carried out in the following manner:

$$C = [c_{i,m}] \in \mathbb{R}^{I\times M}$$

$$XC_{i,*} = [g_1(X_{i,*}^T)g_2(X_{i,*}^T) \dots g_M(X_{i,*}^T)] \tag{7.6}$$

The subsequent step involves determining the members of the set of global best positions, denoted as $\boldsymbol{B}(gbest)$, and then adding the current positions of feasible particles to the set of local best positions, represented by $\boldsymbol{D}_i(pbest)$. The *gbest* set comprises non-dominated positions that the swarm has discovered. Each new feasible particle position is compared against this set to ascertain whether it is non-dominated. Similarly, the pbest set associated with each particle functions similarly but keeps track of feasible positions achieved by the particles themselves. The members of these sets consist of ordered pairs of decision and objective function vectors for non-dominated positions, depicted as follows:

$$\boldsymbol{B} = \left\{\boldsymbol{b}_\phi = \left(\boldsymbol{b}_\phi^x, \boldsymbol{b}_\phi^y\right) | \phi \in \mathbb{N}_{>0}^{\leq\Phi}\right\} \tag{7.7}$$

Here, Φ represents the count of non-dominated positions. Similarly, each particle possesses its own non-dominated set, given by:

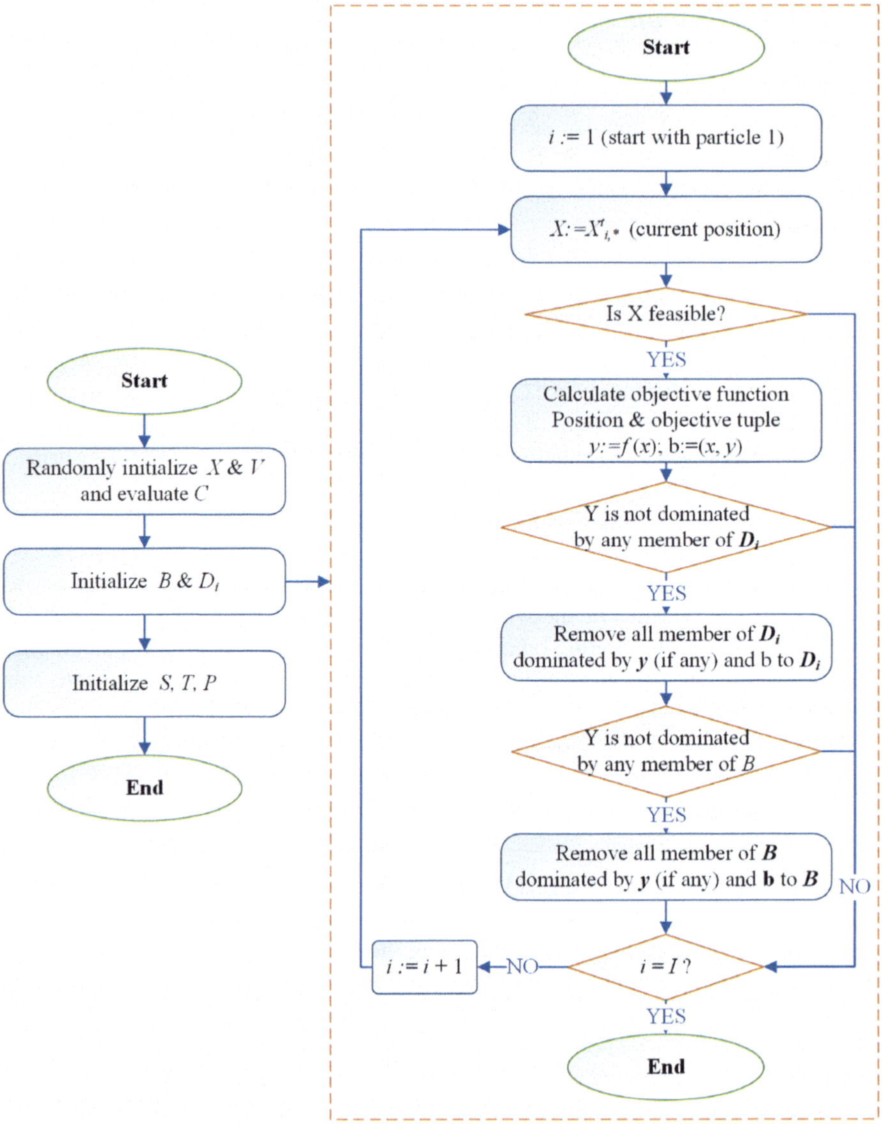

Fig. 7.2 Flowchart of initialization procedure (Hasanoglu and Dolen 2018)

$$\boldsymbol{D}_i = \left\{ \boldsymbol{d}_{i,\omega} = \left(\boldsymbol{d}^x_{i,\omega}, \boldsymbol{d}^y_{i,\omega} \right) | \omega \in \mathbb{N}^{\leq \Omega_i}_{>0} \right\} \tag{7.8}$$

The number of non-dominated positions associated with the i th particle is denoted as Ω_i. If the i th particle is not feasible, its non-dominated set $\boldsymbol{D}_i$ is empty ($\varnothing$), and if there are no feasible particles in the swarm, the global non-dominated set $\boldsymbol{B}$ is also empty (∅). In MOFEPSO, only feasible positions can be part of the *pbest* or *gbest* sets, and the algorithm does not calculate objective functions for infeasible positions.

The initialization of $\boldsymbol{B}$ and $\boldsymbol{D}_i$ requires iterating through all particles. For each feasible particle, its objective function vector is calculated, and its current position is added as the only member to its corresponding *pbest* set. The objective function vector is then compared to the positions in *gbest*. If no positions in *gbest* dominate the objective function vector of the particle, its current position is added to the *gbest* set. Additionally, any positions in *gbest* that are dominated by the current position are removed.

In this context, is said to dominate if and only if:

$$(b_{\vee} \neq b_{\wedge}) \wedge (\leq \forall k \in \mathbb{N}_{>0}^{\leq K} : b_{\vee,k}^{y} \leq b_{\wedge,k}^{y}) \tag{7.9}$$

In MOFEPSO, the movement of infeasible particles is determined by an activated constraint that is chosen at each iteration. In this context, infeasible particles can only move along dimensions where the activated constraint is sensitive. As a result, the matrix $\boldsymbol{S} = [s_{m,n}] \in \mathbb{B}^{M \times N}$ (where *B* represents the Boolean set) is initialized as follows:

$$s_{m,n} = \begin{cases} 1, \textit{constraint } m \textit{ is sensitive to variable } n \\ 0, \textit{otherwise} \end{cases} \tag{7.10}$$

The calculation of constraint sensitivity in MOFEPSO follows a similar approach to the occurrence matrix calculator (OCCURM) in the constraint decomposition technique introduced by Khorshid and Seireg (KhorshidE and Seireg 1999). However, MOFEPSO takes a different approach by temporarily moving initialized particles in one principal direction at a time (a certain percentage of the design variable range x_n^U - x_n^L). This allows the algorithm to identify sensitivities that may exist in specific regions of the search domain. Once the sensitivities are obtained during the initialization phase, the *S* matrix is fixed.

The selection of the activated constraint (AC) depends on constraint priorities and the number of selection recurrences for each constraint. Although these values are updated after each particle flight, they need to be properly initialized. Therefore, the AC selection count matrix $T = [t_{i,m}] \in N^{I \times M}$ is initialized as a zero matrix (as no constraint is initially selected as the AC for any particle). Meanwhile, the constraint priority vector $p = [p_1, p_2, ..., p_M]^T \in N^{I \times M}$ is calculated by counting the number of particles violating each constraint. The constraint violated by the maximum number of particles is assigned the highest priority.

$$p_m = \sum_{i=1}^{I} z_{i,m},\, z_{i,m} = \begin{cases} 1, c_{i,m} > 0 \\ 0, \textit{otherwise} \end{cases} \tag{7.11}$$

Considering the priority of the *m* th constraint as pm. It's important to observe that if $c_{i,m} > 0$, it signifies the violation of that specific constraint.

7.4.2 Infeasible Particle Behavior

Distinct behaviors are exhibited by feasible and infeasible particles within MOFEPSO. This portion delineates the actions executed by particles that are infeasible.

7.4.2.1 Selection of the Activated Constraint and the Global Guide

The movement of particles classified as infeasible is determined by the chosen AC for each particle, which is established at the start of each iteration. The selection of AC is contingent upon the prevailing priorities of constraints and the frequency of constraint selections for each individual particle. The particle opts for the constraint with the highest priority among those that have been chosen the least as the AC. Consequently, for the i th particle, the AC (represented as "a") is chosen in a manner such that

$$a \in R \wedge p_a = \max_{m \in R}(p_m) \tag{7.12}$$

where

$$R = \left\{ m \in \mathbb{N} | t_{i,m} = \min_{r \in R^*}\left(t_{i,r}\right) \right\} \tag{7.13}$$

$$R^* = \left\{ r \in \mathbb{N}_{>N}^{\leq M} | c_{i,r} > 0 \right\} \tag{7.14}$$

In this context, R^* represents the collection of constraints that the ith particle is in breach of, while R signifies the subset of constraints within R^* that have been encountered the fewest times. Following the AC selection, the update to T is as follows:

$$t_{i,a} = t_{i,a} + 1 \tag{7.15}$$

This method of selection guarantees that no high-priority constraint receives excessive focus, maintaining the swarm's diversity. Once the AC is chosen, an infeasible particle can adopt a global guide to direct its trajectory. It's important to note that in MOFEPSO, infeasible particles solely contribute to social influence (without cognitive impact). The guide is selected randomly from the pool of particles that remain compliant with the AC, utilizing

$$J = \{j \in \mathbb{N} > 0 | c_{j,a} \leq 0\} \tag{7.16}$$

Hence, the guide position x^G can be chosen at random from the current positions of particles in set J.

$$x^G = x_{j,*} \tag{7.17}$$

where j represents a randomly chosen element from set $\boldsymbol{J}$.

7.4.2.2 Velocity Update Rules for Infeasible Particles

Assuming that the i th particle provides an infeasible solution, its velocity can be adjusted using

$$v_{i,n} = s_{a,n}\left[Wv_{i,n} + \mathrm{C}_1 r_{1,n}\left(x_n^G - x_{i,n}\right)\right] \tag{7.18}$$

Here, considering W as the inertia factor and C_1 as the social acceleration coefficient, where $r_{1,n} \in [0, 1] \subset R$ are uniformly distributed random numbers. The Boolean sensitivity term $s_{a,n}$ is employed to nullify the velocity term $v_{i,n}$ if the Acceptance Criteria (AC) (a) is insensitive to the decision variable = , effectively halting motion along dimensions where the AC exhibits insensitivity. Subsequent to updating the velocity, particle flight takes place. The process of particle flight remains consistent for both feasible and infeasible particles. Following infeasible particle flight, constraint priorities are adjusted using Eq. (7.15).

7.4.3 Feasible Particle Behavior

In MOFEPSO, feasible particles display a social attraction towards positions in the global best (*gbest*) set (B), while also being cognitively drawn to particles within their own personal best (*pbest*) set (D_i). Consequently, two positions are chosen at random from the *gbest* and *pbest* sets to serve as global (x^G) and personal (x^P) guide positions. To clarify, considering the i th particle as a feasible one, once x^G and x^P are randomly selected from B and D_i correspondingly, the particle's velocity undergoes an update as follows:

$$v_{i,n} = Wv_{i,n} + \mathrm{C}_1 r_{1,n}\left(x_n^G - x_{i,n}\right) + \mathrm{C}_2 r_{2,n}\left(x_n^P - x_{i,n}\right) \tag{7.19}$$

Here, where C_2 represents the cognitive acceleration coefficient, and $r_{1,n}, r_{2,n} \in [0, 1] \subset R$ are random numbers uniformly distributed.

7.4.4 Particle Flight

As the initial step of the MOFEPSO particle flight, a potential new position ($x^* \in \mathbb{R}^{N\times 1}$) is computed, depicted in Fig. 7.3. The components of x^* are determined using the following formulas:

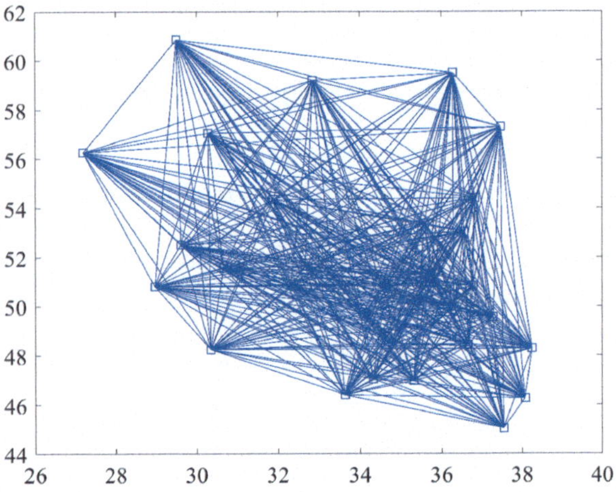

Fig. 7.3 The complete graph of the centers of the provinces of Iran

$$x_n^* = x_{i,n} + v_{i,n} \tag{7.20}$$

Keep in mind that x* is merely a prospective position, as it must fulfill two conditions before being accepted as the particle's new position of significance.

(i) x* must not breach any boundaries set for decision variables.

$$(\nexists n)(x_n^* < x_n^L \vee x_n^* > x_n^U) \tag{7.21}$$

(ii) x* should refrain from violating the constraints that were upheld by the current position of the particle in question (referred to as the i th particle) in the past.

$$(c_{i,m} \leq 0 \wedge c_m^* > 0) \tag{7.22}$$

where c^* represents the constraint vector associated with x^*, formulated as follows:

$$c^* = \left[g_1(\mathrm{x}^*)\ g_1(\mathrm{x}^*) \ldots g_M(\mathrm{x}^*)\right]^T \tag{7.23}$$

MOFEPSO examines the initial criterion mentioned above. If it is met, Eq. (7.27) is assessed to verify the second criterion. However, if the initial criterion is not fulfilled, the decision variable limits are imposed before Eq. (7.27) is taken into account. Similarly, in instances where the second criterion is unmet, the candidate position is adjusted until it aligns with the conditions. It's noteworthy that criterion (ii) ensures that x* maintains compliance with constraints previously adhered to. Consequently, if a particle occupies a feasible position, it is precluded from transitioning to an infeasible position during its trajectory. Once both criteria are satisfied by x*, it is designated as the particle's fresh position, followed by the execution of post-flight operations.

7.4.4.1 Enforcing Decision Variable Limits

In the event that the candidate position x* infringes upon any variable boundaries, a violation rate (w) is calculated to confine the particle within the limits of the decision space.

$$w^L = \max_n \left(\frac{x_n^L - x_n^*}{x_{i,n} - x_n^L} \right) \tag{7.24}$$

$$w^U = \max_n \left(\frac{x_n^* - x_n^U}{x_n^U - x_{i,n}} \right) \tag{7.25}$$

$$W = max(w^L, w^U) \tag{7.26}$$

The candidate position is adjusted utilizing the limit violation rate, leading to the following update:

$$x_n^* := x_{i,n} + \frac{x_n^* - x_{i,n}}{1 + w} \tag{7.27}$$

Similarly, the particle's velocity is reset to zero: $v_{i,n} := 0$.

7.4.4.2 Virtual Boundary Search

The initial phase of the virtual boundary search (VBS) involves designating the virtually feasible (x^+) and virtually infeasible (x^-) boundaries.

$$x^+ := x_i \quad c^+ := c_i \tag{7.28}$$

$$x^- := x^* \quad c^- := c^* \tag{7.29}$$

Subsequently, the rate of infeasibility is computed using the following formula:

$$\max_{m \in Y} \left(\frac{c_m^-}{-c_m^+} \right) \tag{7.30}$$

where Y represents the collection of constraints upheld by the particle's present position, defined as follows:

$$Y = \left\{ m \in \mathbb{N}_{>0}^{\leq M} \right\} | c_{i,m} \leq 0 \} \tag{7.31}$$

The infeasibility rate is utilized to generate a fresh trial position.

$$x^{\tau} := x^{*} + \frac{x^{-} - x^{+}}{1 + q} \tag{7.32}$$

The associated constraint vector c^{τ} is also determined using the following expression:

$$c^{\tau} = \left[g_1(x^{\tau})g_1(x^{\tau})\ldots g_M(x^{\tau})\right]^T \tag{7.33}$$

If the trial position doesn't violate the constraints upheld by the particle's present position, it's deemed virtually feasible according to Eq. (7.26). The concept of virtual feasibility can be reformulated for x^{τ} in the subsequent manner:

$$(\nexists m)(c_{i,m} \leq 0 \wedge c_m^{\tau} > 0) \tag{7.34}$$

Hence, when the condition in (38) holds true, x^{τ} attains virtual feasibility and supersedes x^{+}.

$$x^{+} := x^{\tau} \quad c^{+} := c^{\tau} \tag{7.35}$$

Otherwise, it replaces x^{-}

$$x^{-} := x^{\tau} \quad c^{-} := c^{\tau} \tag{7.36}$$

The iterative process persists by recalculating the infeasibility rate as defined in Eq. (7.34), until the subsequent VBS termination criterion is met.

$$(\forall n)\left(\frac{x_n^{\tau} - x_n^{e}}{x_n^{U} - x_n^{L}} < x^{tol}\right) \tag{7.37}$$

where x^{tol} represents the tolerance, and xe signifies the edge that is nearer to x^{τ}, defined as follows:

$$x^{-} := x^{\tau} \quad c^{-} := c^{\tau} \tag{7.38}$$

Ultimately, the candidate position is adjusted to align with the virtually feasible edge identified through the VBS, resulting in the following update:

$$x^{*} := x^{+} \quad c^{*} := c^{+} \tag{7.39}$$

7.4.5 Post-Flight Operations

The revised candidate position is adopted as the particle's new position, assigning $x_i := x^*$ and$c_i := c^*$. Additional operations are executed after flight, contingent on the particle's pre-flight and post-flight feasibilities. If the particle's prior position was infeasible, Eq. (7.15) is employed to update constraint priorities. Furthermore, when the new position is feasible, adjustments are made to the objective function (OF) vector $[y^* := f(x^*)]$ for both personal best (*pbest*) and global best (*gbest*):

(1) If y^* is not dominated by any element in the set, the pair $b^* = (x^*, y^*)$ is appended to the set.
(2) Any elements in the set dominated by y^* are removed from the set.

7.5 Results and Discussion on Case Studies

In this study, to solve the multi-objective traveling salesman problem, the multi-objective particle swarm optimization algorithm has been improved. To model the problem, the adjacency matrix is defined and the complete graph is used to implement the problem model. The objective functions are defined as distance, traffic, facility, and safety during the round trip. Then the algorithm developed for the discrete mode is implemented on the input data (graph). NSGA-II (Deb et al. 2002) and MOPSO (Lalwani et al. 2013) have been used to compare and evaluate the obtained results. The input data includes the latitude and longitude coordinates of 31 cities related to the centers of the provinces of Iran (Fig. 7.3). Matrices related to objective functions are defined. For example, each row of the distance matrix represents the length of the corresponding edge, and each row of the traffic matrix represents the traffic rate of the corresponding edge.

Finally, the generalized algorithm has been implemented on 31 cities from the centers of the provinces of Iran. Also, results of the FEMOPSO are compared with the NSGA-II and MOPSO. Figure 7.4 shows the routes obtained from the considered algorithms. As can be seen, the route obtained by FEMOPSO algorithm is more optimal than other algorithms.

The average run time of multi-objective algorithms in different implementations for the TSP problem for NSGA-II, MOPSO, and MOFEPSO, were 1.4317s, 0.863s, and 0.628s in 100 iterations, respectively. According to results, the run time of the MOFEPSO algorithm is less than that of the NSGA-II and MOPSO. To determine the convergence process of the algorithms, the obtained optimal values of each iteration are captured. When the variation of the solution in each iteration with the specified number of previous iterations is negligible, it means that the algorithm has converged to the solution. According to Fig. 7.5, the MOFEPSO has converged after in lower iterations from the total of 100 iterations, in comparison to its rivals.

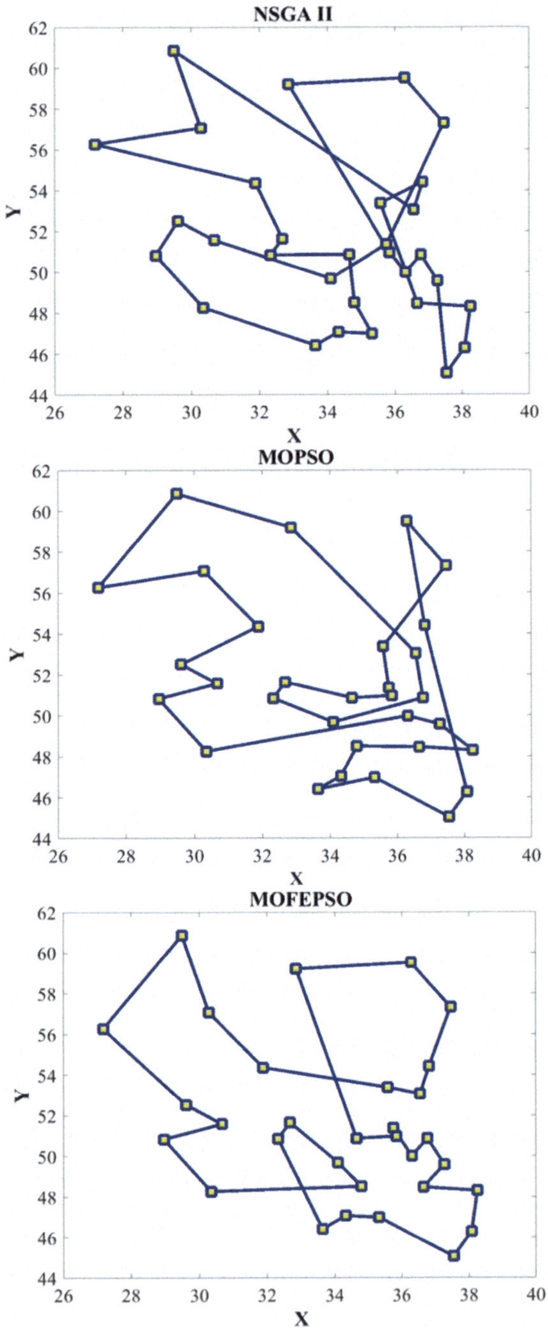

Fig. 7.4 Tendons arrangement at the end and middle sections

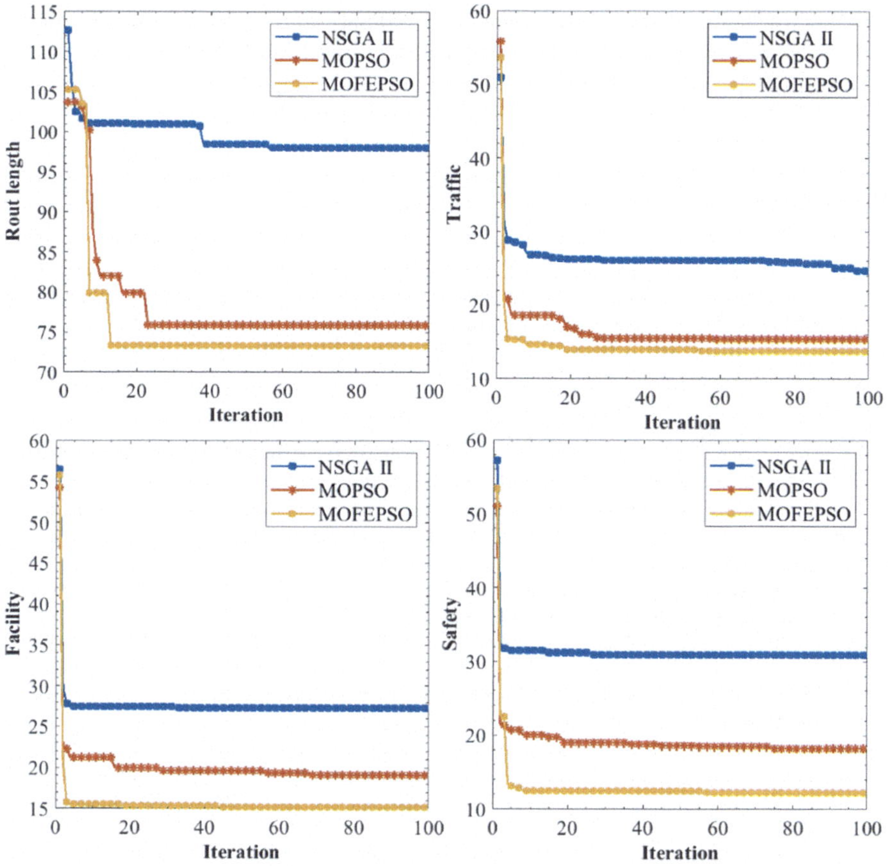

Fig. 7.5 Convergence of algorithms in each objective

7.6 Conclusions

This research has successfully addressed the complex challenge of multi-objective optimization in the context of the Traveling Salesman Problem (TSP) within the cities of Iran. By introducing the Feasibility-Enhanced Particle Swarm Optimization (FEPSO) algorithm, which integrates a feasibility-based approach into the traditional Particle Swarm Optimization (PSO), the study aimed to simultaneously minimize the length of the route and traffic congestion, etc. NSGA-II and MOPSO have been used to evaluate and compare the developed algorithm. To compare algorithms, convergence criteria, algorithm's run time, and the obtained final path have been utilized. According to the results, the MOFEPSO converges to the optimal solution faster. This means that it finds the optimal solution in less time and a lower number of iterations. The answer obtained from MOFEPSO is more accurate and the solution has a minimum value; That is, the MOFEPSO algorithm has found a shorter path with less traffic in comparison with the NSGA-II and MOPSO algorithms. From the findings of this research, it can be concluded that the MOFEPSO is preferable for the traveling salesman problem considering the conflicting objectives.

References

Ashtari P, Karami R, Farahmand-Tabar S (2021) Optimum geometrical pattern and design of real-size diagrid structures using accelerated fuzzy-genetic algorithm with bilinear membership function. Appl Soft Comput 110:107646. https://doi.org/10.1016/j.asoc.2021.107646

Applegate DL, Bixby RE, Chvatal V, Cook WJ (2006) The traveling salesman problem: a computational study. Princeton University Press

Bellman R (1958) On a routing problem. Q Appl Math 16(1):87–90

Bolaños R, Echeverry M, Escobar J (2015) A multiobjective non-dominated sorting genetic algorithm (NSGA-II) for the multiple traveling salesman problem. Dec Sci Lett 4(4):559–568

Clerc M (2004) Discrete particle swarm optimization, illustrated by the traveling salesman problem. In New optimization techniques in engineering (pp. 219–239). Springer, Berlin, Heidelberg

Coello CAC, Lamont GB, Van Veldhuizen DA (2007) Evolutionary algorithms for solving multi-objective problems (Vol. 5, pp. 79–104). New York: Springer

Deb K, Pratap A, Agarwal S, Meyarivan TAMT (2002) A fast and elitist multiobjective genetic algorithm: NSGA-II. IEEE Trans Evol Comput 6(2):182–197

Duman E, Or I (2004) Precedence constrained TSP arising in printed circuit board assembly. Int J Prod Res 42(1):67–78

Du Ke-Lin SM (2016) Search and optimization by metaheuristics, techniques and algorithms inspired by nature

Farahmand-Tabar S, Abdollahi F, Fatemi M (2023). Robust conjugate gradient methods for non-smooth convex optimization and image processing problems. In: Kulkarni, A.J., Gandomi, A.H. (eds) Handbook of Formal Optimization. Springer, Singapore. https://doi.org/10.1007/978-981-19-8851-6_42-1

Farahmand-Tabar S, Ashtari P (2020) Simultaneous size and topology optimization of 3D outrigger-braced tall buildings with inclined belt truss using genetic algorithm. Struct Design Tall Spec Build 29(13):e1776. https://doi.org/10.1002/tal.1776

Farahmand-Tabar S (2023a) Genetic algorithm and accelerating fuzzification for optimum sizing and topology design of real-size tall building systems. In: Dey N (eds) Applied Genetic Algorithm and Its Variants. Springer Tracts in Nature-Inspired Computing. Springer, Singapore. https://doi.org/10.1007/978-981-99-3428-7_9

Farahmand-Tabar S, Ashtari P (2023). Bilinear Fuzzy Genetic algorithm and its application on the optimum design of steel structures with semi-rigid connections. In: Kulkarni AJ, Gandomi AH (eds) Handbook of Formal Optimization. Springer, Singapore

Farahmand-Tabar S, Shirgir S (2023b) Opposed pheromone ant colony optimization for property identification of nonlinear structures. In: Dey N (eds) Applications of Ant Colony Optimization and its Variants. Springer, Singapore

Farahmand-Tabar S, Shirgir S (2023a) Antlion-facing ant colony optimization in parameter identification of the MR damper as a semi-active control device. In: Dey N (eds) Applications of ant colony optimization and its variants. Springer, Singapore

Farahmand-Tabar S (2023b) Frequency-based optimization of truss dome structures using ant colony optimization (ACOR) with multi-trail pheromone memory. In: Dey N (eds) Applications of Ant Colony Optimization and its Variants. Springer, Singapore

Farahmand-Tabar S, Shirgir S (2023) Synergistic collaboration of motion-based metaheuristics for the strength prediction of cement-based mortar materials using TSK model. In: Kulkarni AJ,

Gandomi AH (eds) Handbook of Formal Optimization. Springer, Singapore. https://doi.org/10.1007/978-981-19-8851-6_43-1

Farahmand-Tabar S, Shirgir S (2023) Incorporating Nelder mead simplex as an accelerating operator to improve the performance of metaheuristics in nonlinear system identification. In: Kulkarni AJ, Gandomi AH (eds) Handbook of Formal Optimization. Springer, Singapore. https://doi.org/10.1007/978-981-19-8851-6_39-1

Farahmand-Tabar S, Shirgir S (2023) Boosting the efficiency of metaheuristics through opposition-based learning in optimum locating of control systems in tall buildings. In: Kulkarni AJ, Gandomi AH (eds) Handbook of Formal Optimization. Springer, Singapore. https://doi.org/10.1007/978-981-19-8851-6_37-1

Farahmand-Tabar S, Shirgir S (2023) Positron-enabled atomic orbital search algorithm for improved reliability-based design optimization. In: Kulkarni AJ, Gandomi AH (eds) Handbook of Formal Optimization. Springer, Singapore. https://doi.org/10.1007/978-981-19-8851-6_44-1

Farahmand-Tabar S, Babaei M (2023) Memory-assisted adaptive multiverse optimizer and its application in structural shape and size optimization. Soft Comput. https://doi.org/10.1007/s00500-023-08349-9

Farahmand-Tabar S, Sadrekarimi N (2023) Overcoming constraints: the critical role of penalty functions as constraint handling methods in structural optimization. In: Kulkarni AJ, Gandomi AH (eds) Handbook of Formal Optimization. Springer, Singapore. https://doi.org/10.1007/978-981-19-8851-6_40-1

Farahmand-Tabar S, Rashid TA (2023) Steel plate fault detection using the fitness dependent optimizer and neural networks. In: Kulkarni AJ, Gandomi AH (eds) Handbook of Formal Optimization. Springer, Singapore. https://doi.org/10.1007/978-981-19-8851-6_41-1

Farahmand-Tabar S, Ashtari P, Babaei M (2023) Dynamic intelligence of self-organized map in the frequency-based optimum design of structures. In: Kulkarni AJ, Gandomi AH (eds) Handbook of Formal Optimization. Springer, Singapore. https://doi.org/10.1007/978-981-19-8851-6_45-1

Farahmand-Tabar S (2023c) Memory-driven metaheuristics: improving optimization performance. In: Kulkarni AJ, Gandomi AH (eds) Handbook of Formal Optimization. Springer, Singapore. https://doi.org/10.1007/978-981-19-8851-6_38-1

Farahmand-Tabar S, Ashtari P, Babaei M (2024) Gaussian cross-entropy and organizing intelligence for design optimization of the outrigger system with inclined belt truss in real-size tall buildings. Probab Eng Mech 76:103616. https://doi.org/10.1016/j.probengmech.2024.103616

Farahmand-Tabar S, Shirgir S (2024) Multi-objective adaptive guided differential evolution for passively controlled structures equipped with a tunned mass damper. In: Dey N (eds) Applied Multi-objective Optimization. Springer Tracts in Nature-Inspired Computing. Springer, Singapore. https://doi.org/10.1007/978-981-97-0353-1_3

Farahmand-Tabar S (2024) Multi-objective lichtenberg algorithm for the optimum design of truss structures. In: Dey, N. (eds) Applied multi-objective optimization. Springer tracts in nature-inspired computing. Springer, Singapore. https://doi.org/10.1007/978-981-97-0353-1_5

Farahmand-Tabar S, Afrasyabi P (2024) Multi-modal routing in urban transportation network using multi-objective quantum particle swarm optimization. In: Dey N (eds) Applied multi-objective optimization. springer tracts in nature-inspired computing. Springer, Singapore. https://doi.org/10.1007/978-981-97-0353-1_7

Fang L, Chen P, Liu S (2007) Particle swarm optimization with simulated annealing for TSP. In Proceedings of the 6th WSEAS International Conference on Artificial Intelligence, Knowledge Engineering and Data Bases (pp. 206–210)

Gutin G, Yeo A (2001) TSP tour domination and Hamilton cycle decompositions of regular digraphs. Oper Res Lett 28(3):107–111

Hasanoglu MS, Dolen M (2018) Multi-objective feasibility enhanced particle swarm optimization. Eng Opt. https://doi.org/10.1080/0305215X.2018.1431232

Hendtlass T (2003) Preserving diversity in particle swarm optimisation. In International Conference on Industrial, Engineering and Other Applications of Applied Intelligent Systems (pp. 31–40). Springer, Berlin, Heidelberg

Jozefowiez N, Glover F, Laguna M (2008) Multi-objective meta-heuristics for the traveling salesman problem with profits. J Mathem Model Algor 7(2):177–195

Jourdan L, Basseur M, Talbi EG (2009) Hybridizing exact methods and metaheuristics: a taxonomy. Eur J Oper Res 199(3):620–629

Kaveh M, Kaveh M, Mesgari MS, Paland RS (2020a) Multiple criteria decision-making for hospital location-allocation based on improved genetic algorithm. Appl Geom, 1–16

Kaveh M, Mesgari MS (2019a) Improved biogeography-based optimization using migration process adjustment: an approach for location-allocation of ambulances. Comput Ind Eng 135:800–813

Kaveh M, Mesgari MS (2019b) Hospital site selection using hybrid PSO algorithm-case study: district 2 of Tehran. Sci Res Quart Geogr Data (SEPEHR) 28(111):7–22

Kaveh M, Khishe M, Mosavi MR (2019) Design and implementation of a neighborhood search biogeography-based optimization trainer for classifying sonar dataset using multi-layer perceptron neural network. Analog Integr Circ Sig Process 100(2):405–428

Kaveh M, Mesgari MS, Khosravi A (2020) Solving the local positioning problem using a four-layer artificial neural network. Eng J Geosp Inform Technol 7(4):21–40

Kennedy J, Eberhart RC (1997) A discrete binary version of the particle swarm algorithm. In 1997 IEEE International conference on systems, man, and cybernetics. Computational cybernetics and simulation (Vol. 5, pp. 4104–4108). IEEE

Khishe M, Mosavi MR, Kaveh M (2017) Improved migration models of biogeography-based optimization for sonar dataset classification by using neural network. Appl Acoust 118:15–29

Khorshid EE, Seireg A (1999) Discrete nonlinear optimisation by constraint decomposition and designer

Lalwani S, Singhal S, Kumar R, Gupta N (2013) A comprehensive survey: applications of multi-objective particle swarm optimization (MOPSO) algorithm. Trans Comb 2(1):39–101

Lotfy A, Kaveh M, Mosavi MR, Rahmati AR (2020) An enhanced fuzzy controller based on improved genetic algorithm for speed control of DC motors. Anal Integr Circ Sign Proc, 1–15

Mosavi MR, Kaveh M, Khishe M (2016a) Sonar data set classification using MLP neural network trained by non-linear migration rates BBO. In The fourth Iranian conference on engineering electromagnetic (ICEEM 2016) (pp 1–5)

Mosavi MR, Kaveh M, Khishe M, Aghababaee M (2016b) Design and implementation a sonar data set classifier by using MLP NN trained by improved biogeography-based optimization. In proceedings of the second National Conference on marine technology (pp 1–6)

Mousavi SM, Kaveh M, Khishe M (2016) Sonar data set classification using MLP neural network trained by modified biogeography-based optimization. Iranian J Mar Sci Technol 20(78):65–74

Pires DF, Antunes CH, Martins AG (2012) NSGA-II with local search for a multi-objective reactive power compensation problem. Int J Electr Power Energy Syst 43(1):313–324

Shi Y, Eberhart R (1998) A modified particle swarm optimizer. In 1998 IEEE international conference on evolutionary computation proceedings. IEEE world congress on computational intelligence (Cat. No. 98TH8360) (pp. 69–73). IEEE

Shirgir S, Farahmand-Tabar S, Aghabeigi P (2023) Optimum design of real-size reinforced concrete bridge via charged system search algorithm trained by nelder-mead simplex, Expert systems with applications, 121815. https://doi.org/10.1016/j.eswa.2023.121815

Shuai Y, Yunfeng S, Kai Z (2019) An effective method for solving multiple travelling salesman problem based on NSGA-II. Syst Sci Cont Eng 7(2):108–116

Wang KP, Huang L, Zhou CG, Pang W (2003) Particle swarm optimization for traveling salesman problem. In Proceedings of the 2003 international conference on machine learning and cybernetics (IEEE cat. no. 03ex693) (Vol 3, pp 1583–1585). IEEE

Chapter 8
Drone Path-Planning Leveraging Cohort Intelligence Algorithm

Aryan Bhusari and Anand J. Kulkarni

Abstract In typical combinatorial problems such as Travelling Salesman problem (TSP) or Knapsack problems, it is challenging to calculate the optimal configuration of solutions using brute-force search due to their high computational complexity. Therefore, while dealing with many such problems it is common to employ techniques that prune the obvious inefficient solutions and reduce the search space for the agents involved. In this paper, we address a drone path-planning problem, aimed at surveying the maximum area of a given field within the battery constraints of individual drones. The methodology devised in this article involves a combination of widely recognized algorithms such as backtracking and the greedy algorithm along with a socio-inspired optimization algorithm: Cohort Intelligence. The proposed architecture bases itself on aforementioned algorithms as well as a reward optimization technique that incentivizes of unvisited areas of the field.

Keywords Combinatorial optimization · Cohort intelligence · Real-world problems

8.1 Introduction

A metaheuristic is a higher-level procedure or heuristic that is designed to identify, create, fine-tune, or choose a heuristic that may offer a sufficiently good solution to an optimization problem or a machine learning problem. This is especially valuable in scenarios with incomplete or imperfect information or constrained computational capacity. Examples of popular metaheuristic algorithms, such as Genetic Algorithms (Holland 1992), Simulated Annealing (Aarts et al. 2005), and Particle Swarm Optimization (Shi and Xu 2020), highlight their widespread use and importance in

A. Bhusari (✉) · A. J. Kulkarni
Institute of Artificial Intelligence, MIT World Peace University, Pune, Maharashtra, India
e-mail: 1032201122@mitwpu.edu.in

A. J. Kulkarni
e-mail: anand.j.kulkarni@mitwpu.edu.in

R. Arora and S. Arora (eds.), *Optimization Techniques for Sustainable Environment Under Uncertainty*, Engineering Optimization: Methods and Applications,
https://doi.org/10.1007/978-981-96-0389-3_8

solving complex optimization problems across various domains due to their ability to efficiently explore solution spaces and find near-optimal solutions. Drone path planning involves strategizing optimal routes for unmanned aerial vehicles (UAVs), considering factors like fuel efficiency and safety, to accomplish tasks ranging from surveillance to package delivery and metaheuristics are a good way to handle the path planning.

The objective of this article is to find paths for multiple drones to traverse a rectangular field with defined dimensions, while maintaining their battery level and returning to the initial starting point before their batteries run out. The proposed architecture is robust, in that, it accommodates any number of drones, each capable of having different battery levels. The methodology was modeled similar to the Multiple Travelling Salesman Problem (MTSP). MTSP is a ubiquitous real-world problem where, more than one salesman is involved and each must visit certain number of cities and return to their original starting point. The problem focused on in this study is reminiscent of a MTSP but carries an added constraint of battery level. The general depiction of MTSP can be outlined as follows.

Consider a graph $G = (V, E)$, where V represents the n cities or nodes and E is the set of edges connecting various cities. Generally, the first node is considered as the depot, serving as both the starting and ending point for the salesman. The edges between nodes i and j are represented by non-negative costs c_{ij} which may denote the distance or duration. Each of the salesman starts and ends their journey at the same depot and visit a certain set of nodes. The objective while traversing these routes will be to minimize the overall cost incurred when visiting all the cities. Various constraints can be introduced into the general MTSP based on specific applications and needs. This paper, for instance, introduces a battery constraint for drones, leading to a variation of the problem. The problem that is tackled in this paper closely resembles with two variations of MTSP: Vehicle Routing Problem (VRP) and Multiple Unmanned Aerial Vehicles (MUAVs) Path Planning Problem.

The remainder of this study is structured as follows: Sects. 8.1.1 and 8.1.2 cover VRP and MUAV Path Planning Problem. Section 8.2 elaborates on the Cohort Intelligence Algorithm, and details its application in this field and the problem at hand. Following this, Sect. 8.3 presents the methodology employed in this study and Sect. 8.4 follows up with the results obtained. Finally, Sect. 8.5 summarizes the work done, implications and future directions.

8.1.1 Vehicle Routing Problem

The vehicle routing problem (VRP) is a version of MTSP which involves the minimization of travel costs while considering vehicle capacities. It holds great economic relevance in distribution applications, leading to its increased popularity and attention. VRP is a NP-hard problem making it challenging to calculate the optimal solution. As a result, commercial solvers often resort to heuristics due to the scale and frequency of real-world VRP instances they encounter. VRP is an important real-world problem as it has its applications in package delivery, public transportation, etc. and (Bektas 2006) surveys many of its applications and outlines various formulations along with methods that are used to tackle the problem. Generally, VRP can be formulated as follows:

Let G be a graph defined as $G = (V, E)$, where V $= \{1, 2, 3, \ldots,$ n$\}$. Similar to MTSP, the first node is the starting and end point of the vehicles. C is the capacity of any vehicle and in most cases, it is the same for all the agents or vehicles. Each arc connecting nodes i and j has an associated non-negative cost, denoted as c_{ij}, which could represent either travel distance or time required. Generally, the problem is symmetrical where $c_{ij} = c_{ji}$ for any set of $arc(i, j)$. The objectives of a VRP can vary significantly based on its specific application. Some of the most common objectives include:

a. Minimize the number of vehicles
b. Each location must be visited exactly once
c. Minimize the total cost of traversal
d. Maximize a collective profit/score.

8.1.2 MUAV Path Planning Problem

MUAV path planning involves coordinating the movements of multiple drones or a drone fleet efficiently. The focus is on avoiding collisions, maximizing coverage, allocating tasks among drones, and adapting to changes in the environment. The path planning of MUAVs represents a crucial potential application. However, there has been limited published work applying MTSP/VRP specifically to MUAVs. This area of study is important in various fields like surveillance, search and rescue, agriculture, and more. There are a few ways proposed to deal with MUAV path planning, e.g. (Shi and Xu 2020) proposes MUAV path planning as a combinatorial optimization problem and solve the problem using Particle Swarm Optimization heuristic along with local search techniques. However, the proposed approach in this manuscript, uses an emerging socio-inspired algorithm "Cohort Intelligence" (Kulkarni et al. 2017a) and presents its robustness and potential in dealing with similar combinatorial problems.

The Cohort Intelligence algorithm concentrates on a group of candidates, where each candidate enhances its associated attributes by learning from the best behaviors within the cohort. This process mirrors how humans strive to replicate or learn from one another, contributing to the overall improvement of the cohort. The algorithm's functioning is discussed in further detail in the next section.

8.2 Cohort Intelligence Algorithm

Cohort Intelligence was proposed as an algorithm that allows the candidates in an environment to learn from one another in order to enhance their behavior (quality of candidates), ultimately improving the behavior of the entirety of cohort and thereby converging into optimality (Kulkarni et al. 2013). It is a metaheuristic technique that was inspired from the tendency of individuals in a group to learn and compete with each other. Every candidate attempts to follow a certain other candidate and tries to replicate the other candidate's qualities which may result in enhancement of its own qualities and nudge itself toward achieving a goal common to the whole cohort.

To briefly summarize the working of Cohort Intelligence, we can consider a function as follows:

$$f(\mathbf{x}) = f(x_1, x_2, \ldots, x_n)$$

with the goal being of minimizing the function within certain bounds U (upper) and L (lower).

$$L_i \leq x_i \leq U_i, \textit{where } i = 1, 2, \ldots, n$$

Firstly, a cohort with c (c = 1, 2, ..., C) candidates is initialized, and using the objective function, the probabilities associated with each candidate is calculated using the following formula.

$$p^c = \frac{1/f(\mathbf{x}^c)}{\sum_{c=1}^{C} 1/f(\mathbf{x}^c)}, (c = 1, \ldots, C) \quad (8.2.1)$$

Then using roulette wheel, each candidate selects another candidate or chooses itself to follow. The sampling interval (L, U) shrinks or expands accordingly and once a candidate updates its qualities, the process is repeated if the cohort is not saturated or else the process terminates on convergence. Cohort Intelligence (CI) Algorithm has been employed to solve a wide variety of problems such as Sea Cargo Mix Problem (Kulkarni et al. 2017b), Cross-Border shippers' problem (Kulkarni et al. 2017c), discrete and mixed variable engineering problems (Kale and Kulkarni 2018), etc. It has also proven its robustness by solving various standard combinatorial optimization problems such as 0–1 Knapsack problem (Kulkarni et al. 2017d), TSP (Kulkarni et al. 2017e), etc. making it suitable to deal with a problem like the one being considered.

8.3 Methodology

To model the Drone path-planning problem, it is imperative to set up an environment. Consider a field, that is to be surveyed, of dimension $l \times b$ (Fig. 8.1a). The activity of all the drones is confined within the boundaries of the field. The main purpose of the path-planning is to cover maximum area possible with the given batteries of the field to detect landmines. This problem can also be extended to any alien establishment that needs to be surveyed by unmanned vehicles. The objective for each drone is to utilize its sensor and survey the field without ever re-visiting any part of its trajectory and return to the starting point without exhausting its battery. The payload for each drone is assumed to be the same in this problem. Following are the metrics associated with each drone:

$$n \; number \; of \; drones$$

$$D_i \; ith \, drone, i = 1, 2, \ldots., n$$

$$B_i \; Battery \; of \; ith \, drone, i = 1, 2, \ldots., n$$

$$r \; Sensing \; radius \; of \; Di \; i = 1, 2, \ldots., n$$

The detection sensor on every drone is assumed to have a sensing radius of 'r' (Fig. 8.1b). Hence, the field can be split in terms of the number of circles of radius 'r' that can be fit within the boundaries. To model the problem like a TSP, we can consider the centre of each circle as a city or depot that is to be visited by the salesman (drones in this case). The battery level of a drone in this problem, in definition, is

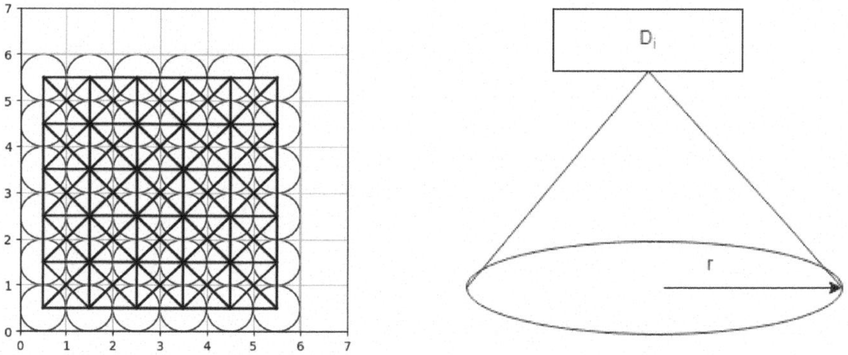

(a) Field of 6 × 6 and r = 0.5, starting point is: (0, 0)

(b) D_i and its sensing radius

Fig. 8.1 Visualization of the problem statement

closer to its mileage as opposed to battery level in terms of Watt. It is defined as the number of cities or points a drone is capable of visiting, until it runs out of battery.

Following steps describe the working of proposed pipeline:

Step 1. The planning assumes a thorough understanding of the terrain, dimensions, nature, and other critical factors concerning the field before the path-planning process commences. The field is assumed to be plain in this study. Using the dimensions of the field and the sensing radius of the drones, the coordinates of checkpoints are calculated. If the radius is not perfectly divisible by the length or breadth of the field, care is taken to consider the residual area as well.

Step 2. Each of the checkpoint or city is assigned a specific reward, that incentivizes them if they are unvisited by any drone. The rewards are determined on the basis of Euclidean distance between the city and starting point. In other words, further the point, higher is the reward associated with it. Because, the proposed methodology employs a universal measure such as Euclidean distance, it can be easily applied to real-life scenarios involving three dimensions. The rewards are assigned to the cities as if it were a quarter of a bullseye (Fig. 8.2), and the starting point is the center. The bullseye method ensures that drones with larger battery capacities are encouraged to explore the outskirts of the field, leaving no area unvisited or at least minimizing unvisited areas. Consequently, this discourages drones with higher battery capacities from lingering near the starting point and compels them to fully explore remote areas, maximizing their potential.

To elaborate further, let us consider a point $P(x, y)$ on a 2-D cartesian plane and origin $O(0, 0)$. The Euclidean distance 'd' point P from O is defined as follows:

$$d = \sqrt{x^2 + y^2} \tag{8.3.1}$$

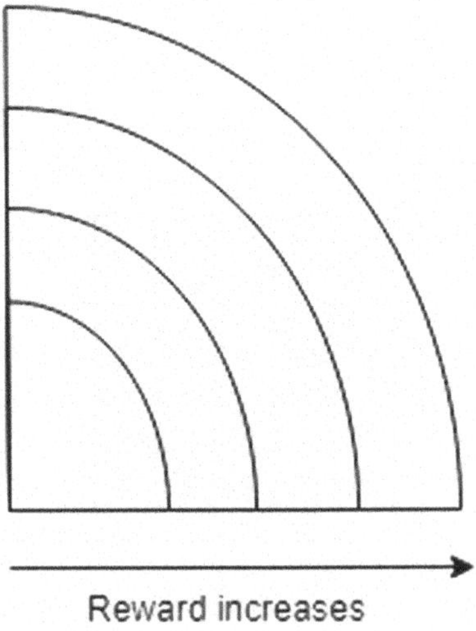

Fig. 8.2 Instance of a quarter of a bullseye

After the calculation, every depot or point in the grid that shares either the x-coordinate or the y-coordinate with P is assigned the same reward. Consequently, the entire field is segmented into regions with specific rewards, which increase as the distance from the origin grows. An illustration of this is presented below.

Step 3. Now the process transitions to the path-planning phase, and for each drone a unique Hamiltonian cycle is constructed through the help of designed python code. The Hamiltonian cycle precisely encompasses the number of cities, ensuring each drone completes the exploration and returns to the starting point without battery depletion issues. This optimization method maximizes the utilization of each drone. To prevent collisions, the algorithm ensures that no two drones visit the same city simultaneously; they visit the same city at different times. An adjacency matrix is maintained for the cities, incorporating travel costs and the reward associated with each city. The reward diminishes to zero upon a drone's initial visit, and subsequent visits incur a fixed deduction from the reward. When two or more adjacent cities near the drone's current location offer the same reward, the decision for the next visit is made at random. If no cities with identical rewards are adjacent, the process proceeds employing a greedy algorithm. Drones prioritize selecting the city with the highest reward during path-planning. In instances where it becomes apparent during path-planning that a drone might exhaust its battery before returning, a back-tracking mechanism is activated. This backtracking mechanism retraces the path to the last optimal city where a different decision could have been made, edging the solution closer to optimality. This strategy discourages repeated visits and promotes efficient exploration. Consequently, individual paths for each drone are computed and consolidated to determine the extent of field coverage, constituting a candidate solution.

Step 4. After generating the initial candidate solution in Step 3, m-1 additional candidate solutions, distinct from both the initial one and each other, are created following the same logic. It is important to note that the steps taken so far might not yield an optimal solution. This is where CI is employed and becomes essential part of the algorithm. The three candidate solutions are treated as a cohort, and specific qualities are attributed to each. These solutions serve as candidates that will learn and compete with one another. Consequently, a total of 'm' candidates is generated to form the cohort. However, for the sake of illustration in this paper, only three candidates are considered as part of the cohort. The concept of treating candidate solutions as agents in a Cohort draws inspiration from the application of CI in solving simple TSP as discussed in Kulkarni et al. (2017e). For each of the candidate the variables are the paths planned for each of the 'n' drones. There were two associated qualities considered in proposed methodology: U_x and O_x, where $x = 1, 2, 3 \ldots, m$. U_x is defined as the total number of unique points visited by all the drones in the field, where a higher value indicates better coverage. O_x, on the other hand, denotes the overlap between drones; it increments by one for each point visited by one or more drones within a candidate solution. In case of O_x, a lower value indicates a superior solution.

Step 5. Leveraging the two associated qualities, roulette wheel selection is utilized to calculate the probabilities associated with each of the candidates. The probability of selection of behavior based on the unique points is calculated for any candidate ‘x’ as follows:

$$P_{Ux}^{x} = \frac{U_x}{U_1 + U_2 + \cdots + U_m}, \; where\, x = 1, 2, \ldots, m \quad (8.3.2)$$

The above roulette wheel formula is employed because the objective is to maximize U_x.

In the case that $U_1 = U_2 = \ldots\ldots = U_m$ for all the candidates, O_x is used as the basis for roulette wheel selection.

$$P_{Ox}^{x} = \frac{1/O_x}{1/O_1 + 1/O_2 + \cdots + 1/O_m}, \; where\, x = 1, 2, \ldots, m \quad (8.3.3)$$

The above roulette wheel formula is employed because the objective is to minimize O_x. Following the calculation of probability of behavior selection, a random number $r \in [0, 1]$ for every candidate $x(x = 1, \ldots, m)$. The roulette wheel approach decides to follow the corresponding behavior of chosen candidate.

Step 6. The learning phase of the process ensues as the probabilities decide the leader and follower. In the case, that the candidate follows itself, no updates are made to it and the qualities are retained from the last iteration. But, in the case that a candidate follows another candidate the variables of the follower attempt to emulate the qualities of the variables of the leader. Here we use a technique called as compartmentalization. The goal of this technique is to compartmentalize the regions of the field to particular drones, which will ultimately reduce overlap and increase the unique points covered, as the drones will rarely cross each other’s path. The best candidate in the cohort will have the greatest compartmentalization and hence larger probability of being the leader.

For instance, consider candidates C_1 and C_2, where C_1 follows C_2. Both constitute the paths for 3 drones D_1, D_2, and D_3. Assuming C_2 is the superior candidate, the unique points covered individually by D_1, D_2, and D_3 of the C_1 and all of the points covered individually by D_1, D_2, and D_3 of the C_2 are combined into 3 sets for the 3 variables and act as the sampling space for C_1 going into the next iteration. The goal is to then evolve C_1 by calculating paths (Hamiltonian cycles) for each drone within the confines of their respective sampling spaces using method used in Step 3. This leads to formation of paths that do not interfere with one another as C_1 followed a better candidate C_2 and hence improves itself by compartmentalizing. Some candidates may end up in a slightly compromised position from where they started, but every iteration improves the cohort as a whole, leading to better solutions. The above steps continue iteratively until the cohort is saturated and all the candidates converge into one form. The process is depicted below (Table 8.1).

Once the paths are found for each drone in iteration 1 for both candidates, the leader and follower are decided based on the roulette wheel probability described in

Table 8.1 Candidates and variables

Candidates	C_1	C_2
Variable (drone paths)	D_1	D_1
	D_2	D_2
	D_3	D_3

Step 5. For the purpose of illustration, let's assume C_1 follows C_2. Sampling space 'S_1' for D_1 drone for all candidates in the next iteration is then defined by following formula:

$$S_1 = U(C_1D_1) + O(C_2D_1) + U(C_2D_1) \tag{8.3.4}$$

$U(C_1D_1)$ represents the unique points covered by C_1D_1 in the latest iteration, and $O(C_2D_1)$ represents the overlapping points covered by C_2D_1 in the latest iteration. To summarize, the sampling space for D_1 in next iteration will be all the points covered by C_2D_1 and unique points covered by C_1D_1. Similarly, the sampling space is calculated for all the drones and the next iteration incorporates those spaces and forms new candidates, that are better than the previous iteration. This fulfils the core goal of CI, i.e. to improve the quality of all the candidates and thereby improving the whole cohort. The specific procedure's comprehensive flowchart is depicted in Fig. 8.3.

8.4 Results

The algorithm was programmed in Python 3.10.11 in Visual Studio Code Editor and the trajectories were plotted and calculated on Windows 11 operating system with Intel core i7 3.80 GHz processor speed with 8 GB RAM. The path-planning algorithm was tested rigorously for different number of drones, with varying dimensions of the field. The CI methodology has shown promise in tackling classic NP-hard combinatorial problems like the TSP. Notably, this capability was showcased within the range of 14 to 29 cities in Kulkarni et al. (2017e). However, the enhanced algorithm not only accommodates multiple salesmen (drones) but also handles more than 29 cities. It is worth noting that as the number of cities to be visited increases, the algorithm's processing time also rises. This is due to the inherent complexity of MTSP, an NP-hard problem that requires exponential time for resolution.

This algorithm offers adaptability to incorporate known obstacles in the drone's field. It can be easily managed by removing the affected city from the adjacency list, rendering it inaccessible to the drones. This step helps prevent collisions by avoiding the obstacle-affected areas during routing. To reiterate the description in the methodology section, an initial path planning is conducted, followed by the application of CI to enhance the resulting plot. It should be emphasized that, in addition to the first plot, two or more plots are generated, collectively forming the

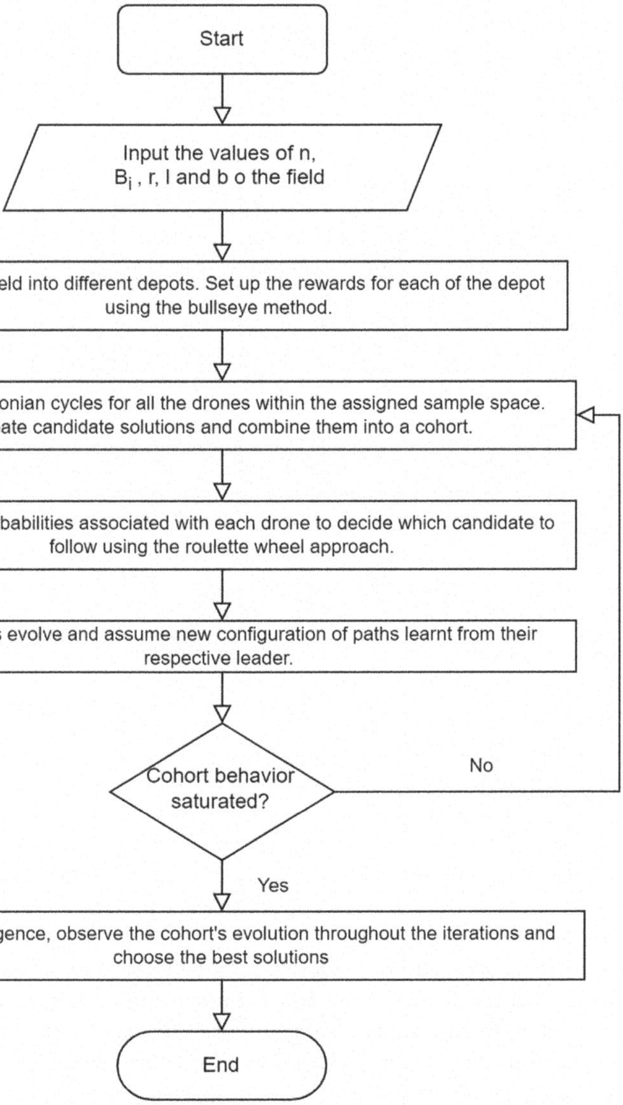

Fig. 8.3 Flowchart of the proposed algorithm

cohort. All candidates within the cohort mutually learn and converge into an improved solution compared to the initial plot.

An illustration depicting the initial solution plots of 3 candidates C_1, C_2 and C_3 and the final converged plot for the plots shown in Fig. 8.1 is provided below.

$$n = 3$$

$$B_1 = 11$$

$$B_2 = 12$$

$$B_3 = 13$$

$$r = 0.5$$

In the Fig. 8.4a the number of unique points and overlap for C_1 is 27 and 6, C_2 is 27 and 5, and C_3 is 27 and 5. And in Fig. 8.4b the number of unique points increases to 30 and overlap reduces to just 3 points. It is worth highlighting that in Fig. 8.4b the drone_3 is able to explore the remotest point on the field unlike Fig. 8.4a, this happens due to the bullseye method, which put the maximum reward on the outer periphery of the field. This shows that proposed methodology is able to effectively compartmentalize the field for different drones and maximize each drone's potential. The combined battery of the drones is 36, and that is why it is highly unlikely that the drones can cover all the points, because the condition is also to return and with minimal overlap. Because the entry points to the starting points are limited to three, as can be seen in Fig. 8.4, overlap among the drones' path will always exist.

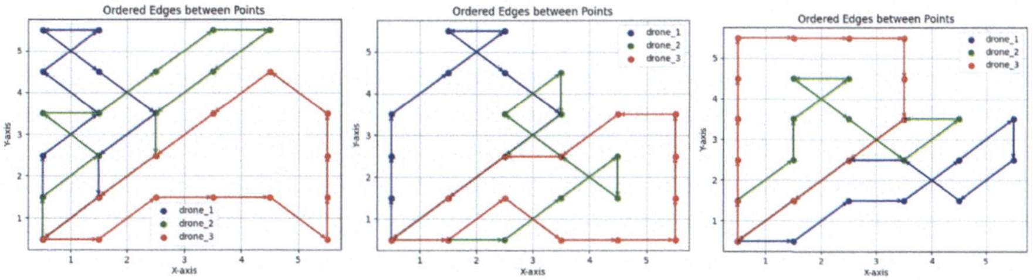

(a) Initial Solutions (3 Candidates)

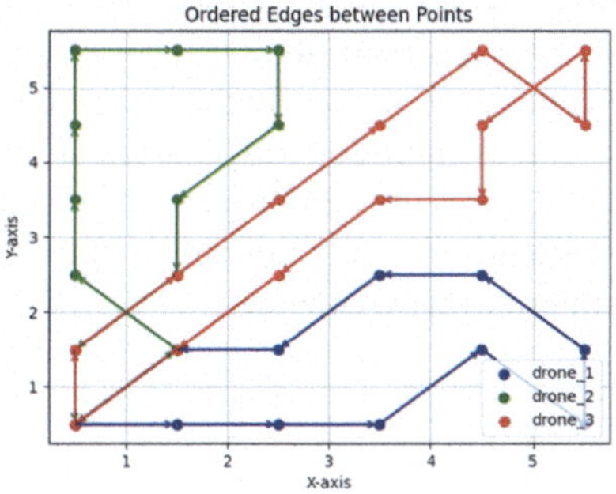

(b) Converged solution with CI

Fig. 8.4 Initial plot versus best solution obtained

In Fig. 8.4a, the count of unique points and overlaps was 27 and 5, respectively. In Fig. 8.4b, the number of unique points increases to 30, and overlaps decrease to just three points. It is worth highlighting that in Fig. 8.4b, drone_3 successfully explores the remotest point on the field, unlike Fig. 8.4a. This is attributed to the bullseye method, which assigns the maximum reward to the outer periphery of the field. This observation underscores the effectiveness of the proposed methodology in compartmentalizing the field for different drones and maximizing each drone's potential. The combined battery capacity of the drones i.e. $B = \sum_{k=1}^{n} B_k$, is 36, making it highly improbable for the drones to cover all points, given the requirement to return with minimal overlap. Since there are only three entry points corresponding to the starting locations (as depicted in Fig. 8.4), overlap among the drones' paths is inevitable.

To recap, the solution depicted in Fig. 8.4a reveals that 6 cities remain unvisited, primarily due to insufficient battery capacity for the drones to explore and return to the starting point with minimal overlap. This illustration serves to demonstrate the effectiveness of the algorithm even under stringent battery constraints. With ample battery capacity, the algorithm ensures coverage of all points by incorporating omitted points into the sampling space of each drone and integrating them into its path, as described in the methodology section.

8.5 Conclusions and Future Directions

This paper presents a comprehensive approach to drone path-planning, merging established algorithms such as greedy algorithm and backtracking with the emerging Cohort Intelligence (CI) algorithm to optimize paths while addressing battery constraints and maximizing field coverage. Through a methodology akin to the MTSP and its resemblance to real-world challenges like the VRP and MUAVs Path Planning Problem, this study showcases its relevance in various domains. The proposed pipeline, emphasizing reward-based planning, and CI-based refinement, exhibits adaptability and scalability beyond classic NP-hard problems. Looking forward, the future scope involves shaping the field diversely to replicate real-world scenarios, incorporating various terrains, and considering alternative starting and ending points for drones to simulate varied operational scenarios. Further advancements may explore adapting paths amidst dynamic environmental changes, integrating machine learning for real-time adaptability, and continually bridging theoretical algorithmic solutions with practical operational challenges.

References

Aarts E, Korst J, Michiels W (2005) Simulated annealing. Search methodologies: introductory tutorials in optimization and decision support techniques, pp 187–210

Bektas T (2006) The multiple traveling salesman problem: an overview of formulations and solution procedures. Omega 34(3):209–219

Holland JH (1992) Genetic algorithms. Sci Am 267(1):66-73

Kulkarni AJ, Krishnasamy G, Abraham A (2017a) Cohort intelligence: a socio-inspired optimization method. Springer International Publishing, Switzerland

Kulkarni AJ, Durugkar IP, Kumar M (2013) Cohort intelligence: a self-supervised learning behavior. In 2013 IEEE international conference on systems, man, and cybernetics (pp 1396–1400). IEEE

Kulkarni AJ, Krishnasamy G, Abraham, A, Kulkarni AJ, Krishnasamy G, Abraham A (2017) Solution to Sea Cargo Mix (SCM) Problem Using Cohort Intelligence Algorithm. Cohort Intelligence: A Socio-inspired Optimization Method, pp 101–115

Kulkarni AJ, Krishnasamy G, Abraham A, Kulkarni AJ, Krishnasamy G, Abraham A (2017) Solution to the Selection of Cross-Border Shippers (SCBS) Problem. Cohort Intelligence: A Socio-inspired Optimization Method, pp 117–129

Kale IR, Kulkarni AJ (2018) Cohort intelligence algorithm for discrete and mixed variable engineering problems. Int J Parallel Emergent Distrib Syst 33(6):627–662

Kulkarni AJ, Krishnasamy G, Abraham A, Kulkarni AJ, Krishnasamy G, Abraham A (2017) Solution to 0–1 knapsack problem using cohort intelligence algorithm. Cohort intelligence: a socio-inspired optimization method, pp 55–74

Kulkarni AJ, Krishnasamy G, Abraham A, Kulkarni AJ, Krishnasamy G, Abraham A (2017) Cohort intelligence for solving travelling salesman problems. Cohort Intelligence: A Socio-inspired Optimization Method, pp75–86

Shi L, Xu S (2020) UAV path planning with QoS constraint in device-to-device 5G networks using particle swarm optimization. IEEE Access 8:137884–137896

Chapter 9
On Finding a Solution of a New Fermatean Fuzzy Transportation Problem: A New Perspective

Aakanksha Singh, Ritu Arora, and Shalini Arora

Abstract Transportation sector is the lifeline of any country's economy which connects the provisioner to the consumer-in-need. Evidently, the words "transportation" and "uncertainties" go hand-in-hand. The transportation cost depends on uncontrollable factors like weather conditions, inflation, toll-stations and other unforeseen circumstances. In the literature, uncertainties in transportation problems have been dealt with by fuzzy sets and their extensions such as intuitionistic fuzzy sets, Pythagorean sets and spherical fuzzy sets. In the proposed work, we have given the solution of a Fermatean fuzzy transportation problem through a new perspective to the already existing Vogel's approximation method. Fermatean fuzzy sets with their arithmetic operations and other details are discussed. The applicability of the proposed work is illustrated using a solved numerical example. The results are also discussed.

Keywords Uncertainty · Transportation problem · Fermatean fuzzy numbers · Vogel's approximation method

MSC Classifications 90B06 · 90C08 · 90C70

A. Singh · S. Arora
Department of Applied Sciences and Humanities, Indira Gandhi Delhi Technical University for Women, Delhi, India
e-mail: aakankshasingh@aryabhattacollege.ac.in

S. Arora
e-mail: shaliniarora@igdtuw.ac.in

A. Singh
Department of Mathematics, Aryabhatta College, University of Delhi, Delhi, India

R. Arora (✉)
Department of Mathematics, Keshav Mahavidyalaya, University of Delhi, Delhi, India
e-mail: ritu.arora@keshav.du.ac.in

R. Arora and S. Arora (eds.), *Optimization Techniques for Sustainable Environment Under Uncertainty*, Engineering Optimization: Methods and Applications,
https://doi.org/10.1007/978-981-96-0389-3_9

9.1 Introduction

Transportation problem (TP) is an optimization problem which deals with transportation of goods from one place to another such that the total transportation cost incurred is minimum and supply-demand constraints are satisfied. The parameters of a TP such as cost, supply and demand in real-life situations are not known exactly/precisely due to uncontrollable factors like road conditions, medical emergencies, fuel costs, volume of the traffic and other unforeseen circumstances. To handle such situations, researchers have been resorting to decision-making in an uncertain environment using fuzzy sets (FSs) and their extensions. FSs introduced by Zadeh (1965) are characterized MGs whose value ranges from zero to one. Atanassov (1986) extended FSs and introduced intuitionistic fuzzy sets (IFSs) defined by the membership grade (MG) and non-membership grade (NMG) conditioned to the relation $(MG) + (NMG) \leq 1$. Since conception, FSs and IFSs have been employed extensively to represent the imprecise parameters of the TP. Kaur and Kumar (2012) solved a fuzzy TP by representing transportation costs by generalized trapezoidal fuzzy numbers (FNs). A new ranking technique based on signed distance was proposed to solve an intuitionistic fuzzy solid TP by Aggarwal and Gupta (2016). Bharati and Singh (2018) solved TP under interval-valued intuitionistic fuzzy environment. Non-linear membership functions were employed to solve a fixed-charge solid TP under intuitionistic fuzzy environment in Chhibber et al. (2021).

Suppose, the decision-maker (DM) assigns the MG and the NMG as 0.5 and 0.8 respectively. In this case, IFSs cannot be used to represent the uncertainty. To handle such cases, Yager (2013) introduced Pythagorean fuzzy sets (PFSs) as an extension of IFSs. PFSs are characterized by MG and NMG s.t. $(MG)^2 + (NMG)^2 \leq 1$. Kumar et al. (2019) worked on Pythagorean fuzzy TP and used the concept of *penalties* to find the initial basic feasible solution (IBFS). Pratihar et al. (2021) handled uncertainty in TP using interval type-2 FS. Recently, Ghosh et al. (2022) considered a multi-objective solid TP with preservation technology using PFS. Saikia et al. (2023) introduced advanced similarity measure for PFSs and worked upon them to solve TPs.

Lately, Fermatean fuzzy sets (FFSs), an extension of PFSs, introduced by Senapati and Yager (2020) are an emerging trend to handle uncertainties as they provide the decision-maker with more values of MG and NMG over which he can exercise his choices. For example, if $(MG, NMG) = (0.8, 0.7)$ then as $0.8 + 0.7 > 1$ and $0.8^2 + 0.7^2 > 1$, the constraint conditions of IFSs and PFSs are not satisfied. However, the relation $0.8^3 + 0.7^3 < 1$ makes the applicability of FFSs more impactful. Sahoo (2021) tackled Fermatean fuzzy TP (FFTP) and introduced a new score function. Jeevaraj (2021) introduced interval-valued FFSs and its applications. Recently, Akram et al. (2022) worked on fractional TP under interval-valued FFSs while Sharma et al. (2023) introduced Fermatean fuzzy programming approach to solve a multi-objective TP. Delving into the literature of different FSs as mentioned above and their applications in TPs, we observed that TP with FFSs is still at its nascent stage and can be explored further to handle uncertainties in transportation sector.

Although Vogel's approximation method (VAM) and its modifications are not new while working with different FNs, the method of calculating the penalty has been majorly subtracting the score values of the two chosen fuzzy transportation costs. The authors in the proposed work give a different perspective to these *penalties* by exploiting the Euclidean distance between them.

We have divided the whole paper into sections. Section 9.2 discusses the definitions of different types of fuzzy sets IFSs, PFSs and FFSs. The set operations and other properties associated with FFSs are also presented. Section 9.3 establishes the FFTP, its solution methodology supported by an illustrative example. Sections 9.4 and 9.5 bring in the discussion and conclusion of the proposed work.

9.2 Preliminaries

Definition 9.1 (*Intuitionistic Fuzzy set* Atanassov (1986)) An IFS defined on a set W $\neq \emptyset$ is an expression form $I_f = \{< w, \alpha_{I_f}(w), \beta_{I_f}(w) >: w \in W\}$, where $\alpha_{I_f}(w), \beta_{I_f}(w) : W \to [0, 1]$ and satisfy the condition $\alpha_{I_f}(w) + \beta_{I_f}(w) \in [0, 1] \forall$ $w \in W$.

Definition 9.2 (*Pythagorean Fuzzy set* Yager (2013)) A PFS on a set W $\neq \emptyset$ is an expression form $P_f = \{< w, \alpha_{P_f}(w), \beta_{P_f}(w) >: w \in W\}$, where $\alpha_{P_f}(w), \beta_{P_f}(w) : W \to [0, 1]$ and satisfy the condition $(\alpha_{P_f}(w))^2 + (\beta_{P_f}(w))^2 \in [0, 1] \forall\ w \in W$. Also, defined is the degree of hesitancy or the degree of indeterminacy of w to P_f as $\pi_{P_f}(w) = \sqrt{1 - (\alpha_{P_f}(w))^2 - (\beta_{P_f}(w))^2}$.

Definition 9.3 (*Fermatean Fuzzy set* Senapati and Yager (2020)) A FFS on a set W $\neq \emptyset$ is an expression form $F_f = \{< w, \alpha_{F_f}(w), \beta_{F_f}(w) >: w \in W\}$, where $\alpha_{F_f}(w), \beta_{F_f}(w) : W \to [0, 1]$ and satisfy the condition $(\alpha_{F_f}(w))^3 + (\beta_{F_f}(w))^3 \in [0, 1] \forall\ w \in W$. Also, defined is the hesitancy degree or the degree of indeterminacy of w to F_f as $\pi_{F_f}(w) = \sqrt[3]{1 - (\alpha_{F_f}(w))^3 - (\beta_{F_f}(w))^3}$.

In the above definitions, $\alpha_A(w), \beta_A(w)$ respectively denote the MG and the NMG of the element w to the set A $(A = I_f, P_f, F_f)$ respectively.
For simplicity, the symbol $F_f = (\alpha_{F_f}, \beta_{F_f})$ will be used for denoting the FFS $F_f = \{< w, \alpha_{F_f}(w), \beta_{F_f}(w) >: w \in W\}$.

Theorem 9.1 (Senapati and Yager (2020)) *The set of Fermatean MGs is larger than the set of Pythagorean MGs and intuitionistic MGs.*

Proof Any point (c_1, c_2) that is an intuitionistic MG is also a Pythagorean MG and a Fermatean MG. For any two numbers $c_1, c_2 \in [0, 1]$, we get $c_1^3 \leq c_1^2 \leq c_1$ and $c_2^3 \leq c_2^2 \leq c_2$. Thus, $c_1 + c_2 \leq 1 \Rightarrow c_1^2 + c_2^2 \leq 1 \Rightarrow c_1^3 + c_2^3 \leq 1$. Therefore, there are Fermatean MGs which are not Pythagorean MGs and intuitionistic MGs. □

Example 9.1 The point $(0.85, 0.6)$ is a Fermatean MG such that $0.85^3 + 0.6^3 \ngtr 1$ but $0.85^2 + 0.6^2 > 1$ and $0.85 + 0.6 > 1$ and hence doesn't qualify to be either Pythagorean MG or intuitionistic MG.

Definition 9.4 (*Set operations on FFS* Senapati and Yager (2020)) Let $F_f^1 = (\alpha_{F_f}^1, \beta_{F_f}^1)$ and $F_f^2 = (\alpha_{F_f}^2, \beta_{F_f}^2)$ be 2 FFSs, then their set operations are defined as

1. $F_f^1 \cap F_f^2 = (min\{\alpha_{F_f}^1, \alpha_{F_f}^2\}, max\{\beta_{F_f}^1, \beta_{F_f}^2\})$;
2. $F_f^1 \cup F_f^2 = (max\{\alpha_{F_f}^1, \alpha_{F_f}^2\}, min\{\beta_{F_f}^1, \beta_{F_f}^2\})$;
3. $(F_f^1)^c = (\beta_{F_f}^1, \alpha_{F_f}^1)$ where c stands for the complement.

Definition 9.5 (*Score and Accuracy function of FFS* Senapati and Yager (2020)) Let $F_f = (\alpha_{F_f}, \beta_{F_f})$ be a FFS. The score and the accuracy function of F_f can be respectively defined as
$\text{Score}(F_f) = (\alpha_{F_f})^3 - (\beta_{F_f})^3$,
$\text{Acc}(F_f) = (\alpha_{F_f})^3 + (\beta_{F_f})^3$.

Clearly, $\text{Score}(F_f) \in [-1, 1]$ and $\text{Acc}(F_f) \in [0, 1]$.

Definition 9.6 (Comparison of two FFSs Senapati and Yager (2020)) Let $F_f^1 = (\alpha_{F_f}^1, \beta_{F_f}^1)$ and $F_f^2 = (\alpha_{F_f}^2, \beta_{F_f}^2)$ be two FFSs with their respective score values as $\text{Score}(F_f^1)$, $\text{Score}(F_f^2)$ and their respective accuracy values as $\text{Acc}(F_f^1)$, $\text{Acc}(F_f^2)$, then

1. If $\text{Score}(F_f^1) < \text{Score}(F_f^2)$, then $F_f^1 < F_f^2$,
2. If $\text{Score}(F_f^1) > \text{Score}(F_f^2)$, then $F_f^1 > F_f^2$,
3. If $\text{Score}(F_f^1) = \text{Score}(F_f^2)$, then

   (a) If $\text{Acc}(F_f^1) < \text{Acc}(F_f^2)$, then $F_f^1 < F_f^2$,
   (b) If $\text{Acc}(F_f^1) > \text{Acc}(F_f^2)$, then $F_f^1 > F_f^2$,
   (c) If $\text{Acc}(F_f^1) = \text{Acc}(F_f^2)$, then $F_f^1 = F_f^2$.

Definition 9.7 (*Euclidean distance of FFSs* Senapati and Yager (2020)) Let $F_f^1 = (\alpha_{F_f}^1, \beta_{F_f}^1)$ and $F_f^2 = (\alpha_{F_f}^2, \beta_{F_f}^2)$ be two FFSs. Then the Euclidean distance between F_f^1 and F_f^2 is given by $d(F_f^1, F_f^2) =$
$\sqrt{\frac{1}{2}[((\alpha_{F_f}^1)^3 - (\alpha_{F_f}^2)^3)^2 + ((\beta_{F_f}^1)^3 - (\beta_{F_f}^2)^3)^2 + ((\pi_{F_f}^1)^3 - (\pi_{F_f}^2)^3)^2]}$.

Example 9.2 Let $F_f^1 = (\alpha_{F_f}^1, \beta_{F_f}^1) = (0.5, 0.4)$ and $F_f^2 = (\alpha_{F_f}^2, \beta_{F_f}^2) = (0.3, 0.2)$. Then,

1. $\text{Score}(F_f^1) = 0.061$ and $\text{Score}(F_f^2) = 0.019$. Thus, $F_f^1 > F_f^2$.
2. $\text{Acc}(F_f^1) = 0.189$ and $\text{Acc}(F_f^2) = 0.035$.
3. $d(F_f^1, F_f^2) = 0.1350$.

9.3 Fermatean Fuzzy Transportation Problem

9.3.1 Mathematical Formulation (Standard TP)

A standard TP is mathematically formulated as

$$Min\ Z = \sum_{i}^{s} \sum_{j}^{d} c_{ij} q_{ij}$$

s.t.

$$\sum_{j}^{d} q_{ij} \leq (sup)_i \text{ where } i = 1, 2, \ldots s, \quad \text{(source constraints)},$$

$$\sum_{i}^{s} q_{ij} \geq (dem)_j \text{ where } j = 1, 2, \ldots d, \quad \text{(demand constraints)},$$

where c_{ij} is the transportation cost associated while delivering one unit of the commodity from source/supply point (SuPt.) i to destination/terminal point (TerPt.) j. In the constraints, $(sup)_i$, $(dem)_j$ are the respective *supply*, *demand* at the ith source point and the jth destination point. The objective is to find how much quantity of the commodity $q_{ij} (\geq 0)$ should be allocated at the (i, j)th location such that the total transportation cost incurred is minimal.
Tabular representation of the above TP can be referred in Table 9.1

Table 9.1 Standard tabular representation of a TP

Destinations → Sources ↓	TerPt.-1	TerPt.-2	…	TerPt.-d	Supply
SuPt.-1	c_{11}	c_{12}	…	c_{1d}	$(sup)_1$
SuPt.-2	c_{21}	c_{22}	…	c_{2d}	$(sup)_2$
⋮	⋮	⋮	⋮	⋮	⋮
SuPt.-s	c_{s1}	c_{s2}	…	c_{sd}	$(sup)_s$
Demand	$(dem)_1$	$(dem)_1$	…	$(dem)_d$	

9.3.2 Mathematical Formulation (Fermatean Fuzzy Transportation Problem (FFTP))

To exhibit the uncertainty associated with the transportation cost in real-life transportation problem such as fuel costs, overtime/night-charges of laborers and safety cost, we model the FFTP with Fermatean fuzzy cost while supplies and demands are crisp numbers. Table 9.2 gives the tabular representation of FFTP in a similar manner as above.

9.3.3 Proposed Methodology

Step-1 Using Definition 9.5, convert all the Fermatean fuzzy costs into crisp costs, i.e. the costs are replaced with their respective score values.
Step-2 Check whether in the given TP, total supply is equal to total demand or not, i.e. is balanced or not. If yes, move to the next step else add dummy source or dummy destination.
Step-3 Identify the least cost $c_{F_f^{ij}}$ and the cost that just exceeds $c_{F_f^{ij}}$, say $c_{F_f^{rs}}$, using the comparison criteria in Definition 9.6. Once identified, calculate the penalty using the Euclidean distance formula given in Definition 9.7, i.e.

$$\begin{aligned} & d(c_{F_f^{ij}}, c_{F_f^{rs}}) \\ & = d((\alpha_{F_f}^{ij}, \beta_{F_f}^{ij}), (\alpha_{F_f}^{rs}, \beta_{F_f}^{rs})) \\ & = \sqrt{\frac{1}{2}[((\alpha_{F_f}^{ij})^3 - (\alpha_{F_f}^{rs})^3)^2 + ((\beta_{F_f}^{ij})^3 - (\beta_{F_f}^{rs})^3)^2 + ((\pi_{F_f}^{ij})^3 - (\pi_{F_f}^{rs})^3)^2]}. \end{aligned}$$

Table 9.2 Tabular representation of a FFTP

Destinations → Sources ↓	TerPt.-1	TerPt.-2	…	TerPt.-d	Supply
SuPt.-1	$c_{F_f^{11}}$[1]	$c_{F_f^{12}}$	…	$c_{F_f^{1d}}$	$(sup)_1$
SuPt.-2	$c_{F_f^{21}}$	$c_{F_f^{22}}$	…	$c_{F_f^{2d}}$	$(sup)_2$
⋮	⋮	⋮	⋮	⋮	⋮
SuPt.-s	$c_{F_f^{s1}}$	$c_{F_f^{s2}}$	…	$c_{F_f^{sd}}$	$(sup)_s$
Demand	$(dem)_1$	$(dem)_2$	…	$(dem)_d$	

[1] $c_{F_f^{11}} = (\alpha_{F_f}^{11}, \beta_{F_f}^{11})$ denotes the Fermatean fuzzy cost when delivering one unit of the commodity from the 1st source/supply point to the 1st destination/terminal point with $MG = \alpha_{F_f}^{11}$ and $NMG = \beta_{F_f}^{11}$

Step-4 Follow the standard VAM procedure to make allocations while calculating the penalties using *Step-3* and obtain the IBFS.
Step-5 Use any of the computing software to obtain the optimal solution, i.e. the value of q_{ij}'s. Obtained q_{ij}'s can be substituted in the objective function to obtain the optimal transportation cost.

9.3.4 Illustrative Example

Let us consider a TP with 3 sources/supply points SuPt.-1, SuPt.-2, SuPt.-3 and 4 destinations/terminal points TerPt.-1, TerPt.-2, TerPt.-3, TerPt.-4 with Fermatean fuzzy costs and crisp supply-demand in Table 9.3.

Solution methodology:
Step-1: Using Definition 9.5, convert all the Fermatean fuzzy costs into crisp numbers (see Table 9.4).

Step-2: Check whether TP is balanced or not.
$\sum_i^3 (sup)_i = \sum_j^4 (dem)_j = 80$. Therefore, it is a balanced TP.
Step-3: Using *Step-3* of Sect. 9.3.3 we calculate all the penalties using Euclidean distance formula. The maximum penalty is identified (0.288). We make an allocation at the cell (2,2). First allotment with penalties calculated is given in Table 9.5.

Table 9.3 FFTP with Fermatean fuzzy cost coefficients

Destinations → Sources ↓	TerPt.-1	TerPt.-2	TerPt.-3	TerPt.-4	Supply
SuPt.-1	(0.5, 0.4)	(0.9, 0.1)	(0.3, 0.2)	(0.6, 0.3)	24
SuPt.-2	(0.6, 0.5)	(0.4, 0.2)	(0.8, 0.3)	(0.7, 0.4)	30
SuPt.-3	(0.9, 0.1)	(0.7, 0.3)	(0.5, 0.3)	(0.8, 0.4)	26
Demand	17	12	23	28	

Table 9.4 Converted FFTP using given score function

Destinations → Sources ↓	TerPt.-1	TerPt.-2	TerPt.-3	TerPt.-4	Supply
SuPt.-1	0.061	0.728	0.019	0.189	24
SuPt.-2	0.091	0.056	0.485	0.279	30
SuPt.-3	0.728	0.316	0.098	0.448	26
Demand	17	12	23	28	

Table 9.5 First allotment with penalties calculated using Euclidean distance formula

Destinations → Sources ↓	TerPt.-1	TerPt.-2		TerPt.-3	TerPt.-4	Supply	Penalties
SuPt.-1	0.061	0.728		0.019	0.189	24	(0.135)
SuPt.-2	0.091	**0.056**	**12**	0.485	0.279	~~30~~ 18	(0.233)
SuPt.-3	0.728	0.316		0.098	0.448	26	(0.218)
Demand	17	~~12~~ 0		23	28		
Penalties	(0.132)	**(0.288)**		(0.109)	(0.148)		

Step-4: Follow the standard VAM procedure to make allocations and calculate the penalties using the Euclidean distance formula.
The IBFS obtained is as follows:
$(S_1, T_4) = q_{14} = 24$,
$(S_2, T_1) = q_{21} = 17$, $(S_2, T_2) = q_{22} = 12$, $(S_2, T_4) = q_{24} = 1$,
$(S_3, T_3) = q_{33} = 23$, $(S_3, T_4) = q_{34} = 3$.
The minimum cost of IBFS obtained is $Z = 24 \times 0.189 + 17 \times 0.091 + 12 \times 0.056 + 1 \times 0.279 + 23 \times 0.098 + 3 \times 0.448 = 10.632$.
Step-5: As the $s + d - 1 = 6$, the solution is degenerate, so next we proceed to test the optimality. To achieve the optimality any computing software can be used.
The optimal solution so obtained is
$(S_1, T_4) = q_{14} = 24$,
$(S_2, T_1) = q_{21} = 17$, $(S_2, T_2) = q_{22} = 12$, $(S_2, T_4) = q_{24} = 1$,
$(S_3, T_3) = q_{33} = 23$, $(S_3, T_4) = q_{34} = 3$.
Optimal cost is $24 \times 0.189 + 17 \times 0.091 + 12 \times 0.056 + 1 \times 0.279 + 23 \times 0.098 + 3 \times 0.448 = 10.632$.
Thus, IBFS = Optimal solution.

9.4 Discussion

As seen in the literature quite often, for finding IBFS of a TP with different types of FNs using VAM, penalties are calculated by subtracting the least and the next to the least score (crisp) values of the transportation costs (Kumar et al. 2019). First allotment with penalties calculated through this technique is given in Table 9.6. In this case the highest penalty obtained is $0.316 - 0.056 = 0.260$.

The IBFS and the optimal solution are the same as obtained in *Step-4* and *Step-5* of Sect. 9.3.4. Thus, we can say that our methodology can also be used to obtain the IBFS while using VAM to solve a TP with Fermatean fuzzy numbers.

Table 9.6 First allotment with penalties calculated by subtracting the score values

Destinations → Sources ↓	TerPt.-1	TerPt.-2		TerPt.-3	TerPt.-4	Supply	Penalties
SuPt.-1	0.061	0.728		0.019	0.189	24	(0.042)
SuPt.-2	0.091	**0.056**	**12**	0.485	0.279	~~30~~ 18	(0.035)
SuPt.-3	0.728	0.316		0.098	0.448	26	(0.218)
Demand	17	~~12~~ 0		23	28		
Penalties	(0.030)	**(0.260)**		(0.079)	(0.090)		

9.5 Conclusion

The proposed work attempts to model the uncertainties in the transportation sector by formulating a FFTP with Fermatean fuzzy transportation costs and crisp supply-demand parameters. Solution methodology revolves around giving a new perspective to the VAM method which is used for finding an IBFS of a TP. As costs are represented by Fermatean fuzzy numbers, for calculating the penalties or the opportunity costs the authors have utilized the concept of Euclidean distance between them. To exhibit the applicability of the proposed work, a FFTP with three sources and four destinations is illustrated. The IBFS using the proposed methodology and an alternate methodology where penalties are calculated using score functions values are also obtained. The results on comparison indicate that proposed method can also be used for finding the IBFS of a FFTP while using VAM. The optimal solution is also obtained using a computing software.

Declarations

Conflict of interest—The authors have no conflict of interest.

Availability of data and materials—The data provided in the paper is hypothetical.

Authors' contributions—Aakanksha Singh—Investigation, Methodology, Software, Writing-Original draft preparation

Ritu Arora—Conceptualization, Visualization, Editing

Shalini Arora—Writing-Reviewing, Supervision

Acknowledgements The authors are grateful to the reviewers for their valuable suggestions, incorporating which has helped us in improving the quality of the chapter to a great extent.

References

Aggarwal S, Gupta C (2016) Solving intuitionistic fuzzy solid transportation problem via new ranking method based on signed distance. Int J Uncertainty, Fuzziness Knowl Based Syst 24(04):483–501. https://doi.org/10.1142/S0218488516500240

Akram M, Shah SMU, Al-Shamiri MMA, Edalatpanah SA (2022) Fractional transportation problem under interval-valued Fermatean fuzzy sets. AIMS Math 7(9):17327–17348

Atanassov KT (1986) Intuitionistic fuzzy sets. Fuzzy Sets Syst 20:87–96

Bharati SK, Singh SR (2018) Transportation problem under interval-valued intuitionistic fuzzy environment. Int J Fuzzy Syst 20:1511–1522

Chhibber D, Bisht DC, Srivastava PK (2021) Pareto-optimal solution for fixed-charge solid transportation problem under intuitionistic fuzzy environment. Appl Soft Comput 107:107368. https://doi.org/10.1016/j.asoc.2021.107368

Ghosh S, Roy SK, Fügenschuh A (2022) The multi-objective solid transportation problem with preservation technology using Pythagorean fuzzy sets. Int J Fuzzy Syst 24(6):2687–2704

Jeevaraj S (2021) Ordering of interval-valued Fermatean fuzzy sets and its applications. Expert Syst Appl 185:115613

Kaur A, Kumar A (2012) A new approach for solving fuzzy transportation problems using generalized trapezoidal fuzzy numbers. Appl Soft Comput 12(3):1201–1213. https://doi.org/10.1016/j.asoc.2011.10.014

Kumar R, Edalatpanah SA, Jha S, Singh R (2019) A Pythagorean fuzzy approach to the transportation problem. Complex Intell Syst 5(2):255–263

Pratihar J, Kumar R, Edalatpanah SA, Dey A (2021) Modified Vogel's approximation method for transportation problem under uncertain environment. Complex Intell Syst 7(1):29–40

Sahoo L (2021) A new score function based Fermatean fuzzy transportation problem. Results Control Optim 4:100040

Saikia B, Dutta P, Talukdar P (2023) An advanced similarity measure for Pythagorean fuzzy sets and its applications in transportation problem. Artif Intell Rev 56(11):12689–12724. https://doi.org/10.1007/s10462-023-10421-7

Senapati T, Yager RR (2020) Fermatean fuzzy sets. J Ambient Intell Humanized Comput 11:663–674

Sharma MK, Kamini Dhaka A, Nandal A, Rosales HG, Monteagudo FEL, Hernández AG, Hoang VT (2023) Fermatean fuzzy programming with new score function: a new methodology to multi-objective transportation problems. Electronics 12(2):277. https://doi.org/10.3390/electronics12020277

Yager RR (2013) Pythagorean fuzzy subsets. In: 2013 joint IFSA world congress and NAFIPS annual meeting (IFSA/NAFIPS). IEEE, pp 57–61

Zadeh LA (1965) Fuzzy sets. Inf Control 8(3):338–353. https://doi.org/10.1016/S0019-9958(65)90241-X

Chapter 10
GrowWise: Machine Learning Driven Crop, Soil and Fertilizer Assistant

Shreyansh Khokale, Priyanshi Kotian, Akshada Ghoderao, Shruti Nare, Prachi Rajarapollu, and Savita Pawar

Abstract In order to ensure the protection of food, sustainable agriculture techniques are essential given the expanding global population and growing demand for food. Efficient crop selection and effective use of fertilizers are essential for maximizing productivity and reducing environmental impact. GrowWise is a web-based tool offered by this observation that uses machine learning techniques to help farmers determine the best crop to plant and the optimal amount of fertilizer to use based on the nutritional composition of the soil. To improve model performance, our approach includes thorough data exploration, visualization, preprocessing, and feature engineering. The use of K-fold cross-validation to provide robustness and eliminate outliers is done, in order to address problems with data quality. Several performance metrics were used to assess the created models: mean squared error was used for soil nutrient composition and accuracy, F1-score, precision, recall, and support was used for crop and fertilizer recommendation. The outcomes show GrowWise's ability to support sustainable farming practices and higher crop yields by offering precise advice on crop selection and fertilizer application. This study shows how machine learning (ML) may be used to create user-friendly decision support systems like GrowWise. These systems have the potential to greatly improve farmers' ability to make decisions, optimize the use of their resources, and ultimately raise agricultural productivity.

Keywords Precision agriculture · Crop selection · Fertilizer recommendation · Machine learning · Ensemble learning · Data preprocessing · Feature engineering · Random forest · XGBoost · SVM · KNN · K-fold cross-validation · Hyperparameter tuning

S. Khokale · P. Kotian · A. Ghoderao · S. Nare · P. Rajarapollu (✉) · S. Pawar
Department of Electronics and Telecommunication Engineering, MIT Academy of Engineering, Pune, Maharashtra, India
e-mail: prrajarapollu@mitaoe.ac.in

R. Arora and S. Arora (eds.), *Optimization Techniques for Sustainable Environment Under Uncertainty*, Engineering Optimization: Methods and Applications,
https://doi.org/10.1007/978-981-96-0389-3_10

10.1 Introduction

In order to maintain food security and support the world's population, agriculture is essential. It is now essential to implement sustainable agriculture practices that maximize crop yields while minimizing environmental effect due to the world's growing food need. The careful selection of crops and the accurate application of fertilizers based on the nitrogen composition of the soil are two important elements determining crop productivity.

Conventional farming methods frequently depend on empirical information and trial-and-error techniques, which can result in poor crop selection and either excessive or inadequate fertilizer application. These methods not only provide poorer yields but also exacerbate greenhouse gas emissions, water pollution, and soil deterioration. Furthermore, it is difficult for farmers to make wise decisions due to the intricate interactions between crop requirements, meteorological circumstances, and soil characteristics.

Precision agriculture is now possible thanks to recent developments in machine learning (ML) and data analytics (Rama Krishna and Amrutha 2020), which make data-driven decision-making processes possible. Large datasets and sophisticated machine learning algorithms can be used to create predictive models that help farmers choose the best crops and apply fertilizer at the right rates (Praynlin et al. 2023).

In this work, introduce GrowWise, a web-based tool that uses machine learning techniques to give farmers advice on crop choices and fertilizer application. The suggested approach makes use of a large dataset that includes details on crop yield, weather patterns, and soil properties (Chana et al. 2023). We prepare the dataset for model training using preprocessing approaches, data visualization, and exploratory data analysis. Outliers are eliminated to improve model performance, and feature engineering is used to extract pertinent characteristics.

For crop categorization and fertilizer recommendation tasks, we compare the effectiveness of many machine learning (ML) methods (Mohammad et al. 2022), such as Random Forest, XGBoost, Support Vector Machines (SVM), and K-Nearest Neighbors (KNN). To integrate the advantages of different models, ensemble learning techniques are also used, such as boosting and voting. To maximize model performance, hyperparameter tuning and K-fold cross-validation are applied (Suchithra and Pai 2020; Bellis Emily et al. 2022; Rohit Kumar et al. 2017).

Several performance metrics are used to assess the created models: mean squared error is used for soil nutrient composition, and accuracy, F1-score, precision, recall, and support is used for crop and fertilizer recommendation (Macharia et al. 2016). The outcomes show GrowWise's ability to support sustainable farming practices and higher crop yields by offering precise advice on crop selection and fertilizer application.

10.2 Problem Statement

Even though agricultural data is readily available, farmers frequently lack the knowledge and understanding needed to make decisions that are appropriate for their particular operational contexts. A comprehensive decision support system that incorporates cutting-edge technology like machine learning (ML) is desperately needed to evaluate intricate agricultural datasets and offer tailored crop selection and fertilizer application suggestions.

In the current lack of such a solution, farmers are exposed to the unknowns and hazards that come with using less-than-ideal farming methods, such as decreased yields, wasted inputs, and environmental damage. In order to close this gap, a cutting-edge platform that uses ML models to mine the abundance of agricultural data accessible and convert it into insights that farmers can use must be created (Priyadharshini et al. 2021; Durai Arumugam et al. 2023).

10.3 Related Work

1. A Novel Ensembling Technique For Crop Recommendation System Using Machine Learning.

A crop recommendation ensemble model, comprising individual base learners like Artificial Neural Network (ANN), Random Forest, Support Vector Machine (SVM), and K-Nearest Neighbors (KNN), is proposed in this paper. When Random Forest and Back-propagation error rates are compared, Random Forest has a lower error rate. The article also covers other methods for identifying soil types and recommending crops, such as decision trees, neural networks, naive bayes, k-means, KNN, SVM, and Bayesian networks. The use of SVM classification, which has the highest accuracy, Decision Trees, Logistic Regression, and SVM classification for crop recommendation are also covered. The effect of KNN, SVM, LMNN, and ENN models on data classification is also discussed in the paper (Rama Krishna and Amrutha 2020).

Overall, for crop recommendation and fertilizer suggestion, the paper uses a combination of machine learning algorithms, including ANN, Random Forest, SVM, KNN, k-means, Decision Tree, Logistic Regression, and SVM.

2. Machine Learning Approach for Crop and Fertilizer Recommendation.

In this paper, a customized recommendation system for crop and fertilizer selection based on particular soil and weather conditions is developed using machine learning algorithms.

The authors go over a number of techniques used in agriculture for precise nutrient predictions and crop selection, including SVM algorithms, photo processing, and bio-inspired frameworks. The importance of feature selection and classification in crop prediction is also covered in this paper, along with the application of deep learning techniques like RNN and LSTM for crop yield prediction (Praynlin et al. 2023).

They present Boruta and RFE feature selection techniques. The study addresses the difficulties in analyzing agricultural data and highlights the use of devices and sensors for agricultural monitoring. The use of deep neural techniques for fake news detection, such as LSTM, is also mentioned by the authors (Zhu et al. 2018).

3. Real-Time Crop Prediction Based on Soil Fertility and Weather Forecast Using IoT and a Machine Learning Algorithm.

The research uses artificial intelligence (AI) and the internet of things (IoT) to predict crops in real time based on weather and soil fertility. Random Forest machine learning algorithm is used in the study's prototype to achieve crop prediction accuracy of approximately 99%. Along with weather forecast data from a weather API, the IoT system gathers input parameters from sensors, including temperature, pH, humidity, phosphorus, and potassium (Chana et al. 2023).

The use of other machine learning algorithms, such as the Decision Tree algorithm, K-Nearest Neighbor (K-NN), Support Vector Machine (SVM), Extreme Learning Machine (ELM), and linear regression, in related studies for enhancing soil health and enabling farmers to select appropriate crops based on rainfall and market crop price, is also mentioned in the paper.

4. A Machine Learning Method for Crop Recommendation

The study proposes a thorough plan for using mobile technologies and machine learning algorithms to enhance productivity in agriculture in rural India. The study aims to provide farmers with an accessible system for forecasting crop yields and improving agricultural techniques with the help of GPS and smartphone applications. After comparing analyses of datasets from Maharashtra and Karnataka, Random Forest Regression is shown to be the most accurate model to predict crop yield among the many machine learning algorithms studied. The research additionally includes a useful yield forecasting system that is connected to a smartphone app, providing direct farmer-to-crop selection support communication (Mohammad et al. 2022). To help farmers plan ahead, the app provides tools for crop selection, accurate production estimates, and suggestions for crop varieties. In order to maximize earnings and enhance agricultural outcomes, the research additionally promotes the timely use of fertilizers based on crop prediction models. Overall, the study highlights how critical it is to use cutting-edge technologies in order to improve agricultural practices and offer data-driven insights, ultimately supporting the sector's sustainable growth and the well-being of rural Indian farmers.

5. Fertilizer use optimization approach: An innovation to increase agricultural profitability for African farmers.

The goal of the Optimization of Fertilizer Recommendations in Africa (OFRA) project was to create crop-specific fertilizer recommendations and agroecological zones (AEZs) for 13 countries in Sub-Saharan Africa (SSA). The project's goal was to assist smallholder farmers in maximizing returns on investment and optimizing production when using fertilizer (Macharia et al. 2016).

For 14 significant crops in Africa, the research created 65 AEZ-specific fertilizer optimization tools (FOTs). Two supplementary tools were created in addition to the FOTs: the nutritional substitution table and the paper tool. A number of variables, including crop value, fertilizer prices, crop response to nutrients, and response curve shape, affected how profitable various combinations were. The strategy sought to increase the productivity and profitability of smallholder farmers in Africa by addressing their poor fertilizer use.

6. Intelligent Crop Recommendation System using Machine Learning

The study applies machine learning methods, notably ensemble models and regression analysis, to intelligent crop recommendation. To recommend crops based on soil data, five distinct algorithms are used: Support Vector Machine, Bagged Tree, Ada boost, Naive Bayes, and Artificial Neural Network. The suggested recommendation system uses CHAID, K-Nearest Neighbor, Naive Bayes, and Random Tree as learners in an ensemble model with a majority voting mechanism. We talk about precision agriculture, which recommends the best crops depending on site-specific factors (Priyadharshini et al. 2021).

To ascertain how profit affects crop prediction, profit analysis is carried out utilizing the yield dataset, market price, standard price, and cultivation costs. In order to help farmers choose crops, the study also takes into account elements like planting season, soil type, and geographic location.

7. An Intelligent Crop Recommendation System using Deep Learning

In order to estimate crop yield, the research article "An Intelligent Crop Recommendation System using Deep Learning" examines important variables such soil nutrients, water bodies, and climate. In order to achieve this, the study processes and extracts insights from data pertaining to soil parameters and crop attributes using a hybrid approach that combines neural network algorithms with image processing techniques. Furthermore, data on water bodies is gathered via satellite remote sensing technology, which is essential for environmental research and climate prediction (Durai Arumugam et al. 2023).

The goal of the project is to improve crop recommendation systems' accuracy and efficiency by using these cutting-edge technology. Neural network algorithms are essential for jobs involving pattern identification and classification, and image processing techniques aid in the extraction of useful data from photographs of crop leaves and soil samples. The researchers' ability to gather information on water bodies through satellite remote sensing adds to a more thorough understanding of the environmental elements influencing crop productivity (Sambandh Bhusan et al. 2022).

8. Crop yield prediction using multi-attribute weighted tree-based support vector machine

The paper utilizes the MAWT-SVM approach for crop yield prediction in agriculture. It involves data collection, preprocessing with PCA and Genetic Algorithm, and model development using support vector machine techniques. Performance evaluation metrics are employed to compare the model with existing methods, highlighting its effectiveness. The study emphasizes the importance of accurate crop yield forecasting for sustainable agriculture and suggests avenues for future research in this field (Rajakumaran et al. 2024).

9. Machine learning in nutrient management: A review

The paper discusses the application of machine learning techniques in nutrient management in agriculture. It reviews various methods and approaches developed over the last decade for estimating fertilizer and nutrient status. The study explores detection and classification techniques, including machine learning algorithms, feature selection, and performance metrics such as accuracy, recall, precision, F-measure, and area-under-the-curve. Additionally, the paper highlights the use of technologies like soil spectroscopy, chemometrics, neural networks, and PLSR in developing predictive models for nutrient management. The integration of data from different sources and the importance of data quality and quantity are emphasized to enhance the accuracy and practical application of machine learning in agricultural nutrient management (Ennaji et al. 2023).

10. Implementation of Machine Learning Algorithms for Crop Recommendation Using Precision Agriculture

The application of machine learning and data mining to agriculture for crop selection based on soil requirements is examined in the study "Implementation of Machine Learning Algorithms for Crop Recommendation Using Precision Agriculture". It discusses the difficulty Indian farmers have choosing the best crop for their soil, which affects output (Akshatha and Shreedhara 2018; Kamilaris and Prenafeta-Boldú 2018). The suggested recommendation system makes use of machine learning techniques including Random tree, K-Nearest Neighbor, and Naive Bayes in an ensemble model using a majority voting technique. The method seeks to deliver precise crop recommendations that are customized to each individual soil condition by taking into account soil-specific characteristics including depth, texture, pH, and water-holding capacity. This will ultimately increase crop production and productivity (Adamchuk Viacheslav et al. 2004).

10.4 Methodology

The GrowWise platform is intended to use machine learning techniques to provide farmers advice on fertilizer application and crop choices. Data collection, exploratory data analysis, data preprocessing, feature engineering, model creation, and evaluation are some of the phases in the approach used in this study (Rajakumaran et al. 2024).

1. Data Collection and Description:

An extensive dataset was assembled from multiple sources, such as government organizations, meteorological departments, and agricultural research institutions.

The dataset includes agricultural yield statistics for several locations, as well as soil parameters (such as pH) and meteorological factors (such as temperature and rainfall). The broad range of crops and soil types included by the dataset guarantees the generalizability of the models that are constructed.

2. Exploratory Data Analysis and Data Preprocessing:

Techniques for exploratory data analysis (EDA) were used to analyze the dataset and find any trends, outliers, and problems with data quality. Correlation matrices and other data visualization tools were used to comprehend the relationships between variables and find possible characteristics for the creation of the model.

To deal with missing values, get rid of duplicates, and guarantee data consistency, preprocessing procedures were carried out. The interquartile range (IQR) approach was used to find and eliminate outliers in order to enhance model performance and lessen the influence of extreme data.

3. Feature Engineering:

In this study, feature engineering was quite important. Utilizing domain expertise, new features were developed, such as nutrient ratios, growing degree days, and soil moisture indices, which captured pertinent data for crop selection and fertilizer recommendations. To convert categorical variables into a format that is appropriate for machine learning algorithms, encoding techniques such as label encoding, target encoding, and one-hot encoding were utilized. To improve the discriminative power of continuous data, domain-specific functions were applied, such as logarithmic transformations for soil pH. By removing the most useful characteristics from the dataset, feature engineering sought to increase the resilience and accuracy of the created models.

4. Model Development and Evaluation:

For the tasks of crop selection and fertilizer recommendation, a number of machine learning algorithms were assessed, including K-Nearest Neighbors (KNN), XGBoost, Random Forest, and Support Vector Machines (SVM). These algorithms were selected because to their track record of success in related agricultural applications and their capacity to manage jobs involving both regression and classification. To enhance performance overall, ensemble learning techniques like gradient boosting and vote classifiers were used to aggregate the advantages of various models. While gradient boosting iteratively trains weak models and combines them to generate a strong predictive model, voting classifiers aggregate the predictions of numerous models. K-fold cross-validation was utilized to assess the generalization performance of the developed models and prevent overfitting. The dataset was split into training and testing sets, and the models were trained and evaluated using different folds of the data. Hyperparameter tuning was performed to optimize the model parameters and achieve the best possible performance. Grid search and random search techniques were employed to explore the hyperparameter space and identify the optimal combination of parameters for each algorithm (Ennaji et al. 2023).

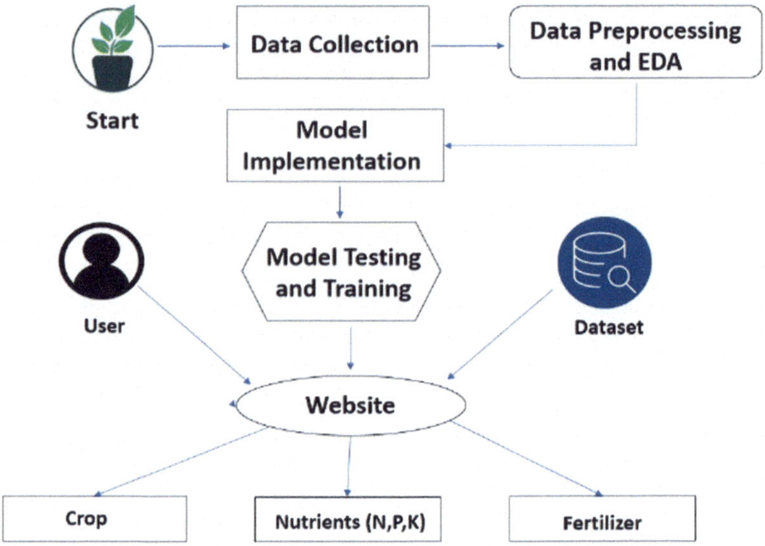

Fig. 10.1 Data flow diagram for the implemented system

The model's performance was assessed using a range of metrics, including as mean squared error (MSE) for fertilizer recommendation tasks (regression issues) and accuracy, F1-score, precision, recall, and support for crop classification tasks. These metrics included information on the models' precision in predicting crop compatibility and recommending the right rates of fertilizer application (Akshatha and Shreedhara 2018). Figures 10.1 and 10.2 gives Data Flow Diagram for the implemented system and model training pipeline.

10.5 Results and Observations

The accuracy comparison of crop suggestions from several machine learning models is displayed in Fig. 10.3.

With an average accuracy of 0.867, all of the models have accuracy levels above 0.80.

At 0.869, Random Forest has the highest accuracy.

Second place goes to XGBoost with 0.867.

The accuracy of KNN, Random Forest + Gradient Boosting, and SVM is approximately 0.85.

The mean squared error (MSE) for several models that are used to forecast nutrient requirements is displayed in the Fig. 10.4. In this graph, a lower mean square error (MSE) indicates greater performance since it indicates that the model's predictions are generally closer to the actual values. XGBoost appears to be the model with the

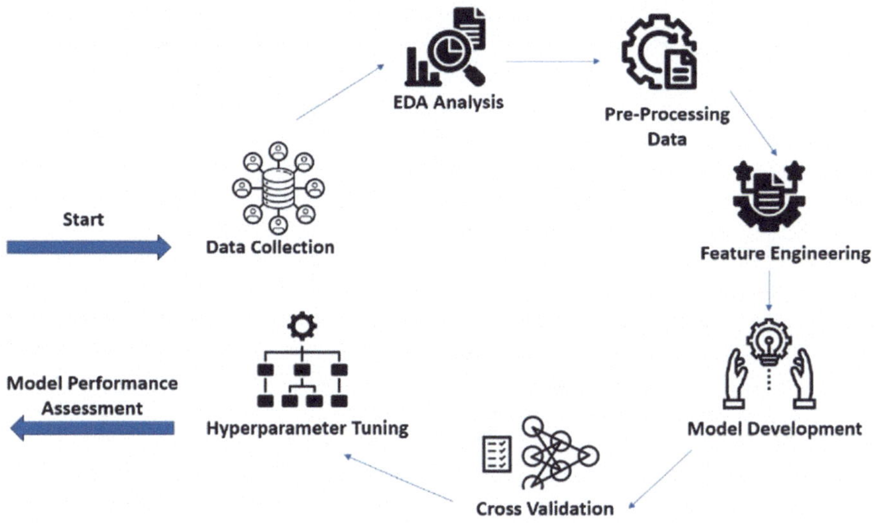

Fig. 10.2 Model training pipeline

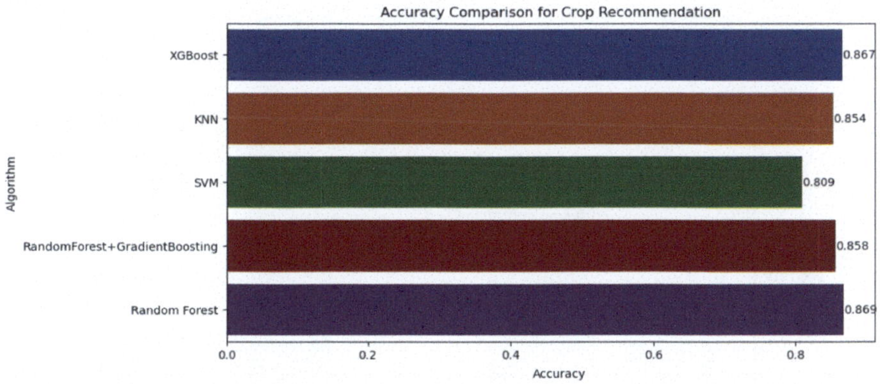

Fig. 10.3 Accuracy comparison for crop recommendation

lowest MSE, followed by Random Forest. This implies that the nutrient requirements predicted by these models are, on average, closest to the actual amounts required. When forecasting the amount of nutrients needed, SVM has the greatest MSE, which indicates that its average errors are the biggest.

The Fig. 10.5 displays the Accuracy Comparison for Fertilizer Recommendation. Out of which XGBoost has the highest accuracy of 0.991 followed by Random Forest with accuracy 0.894.

Variable Correlations: The matrix displays the correlation coefficients between a number of variables that are probably important for crop development and fertilizer needs. These factors include temperature, pH, rainfall, phosphorus (P), potassium

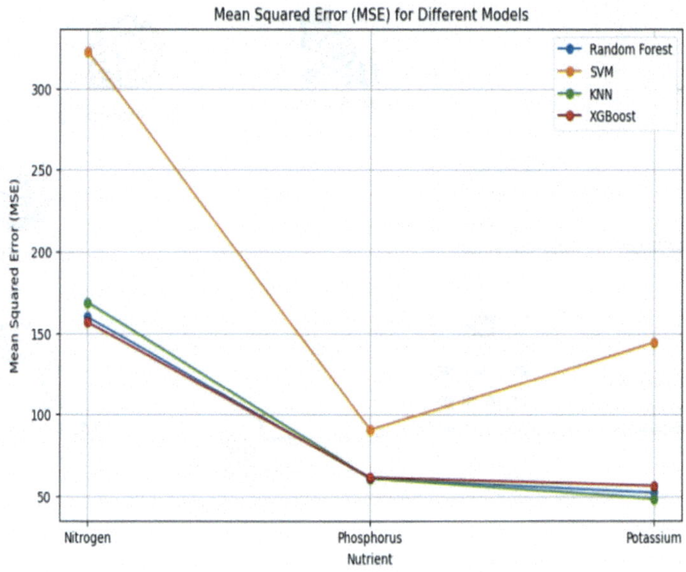

Fig. 10.4 N,P,K optimization

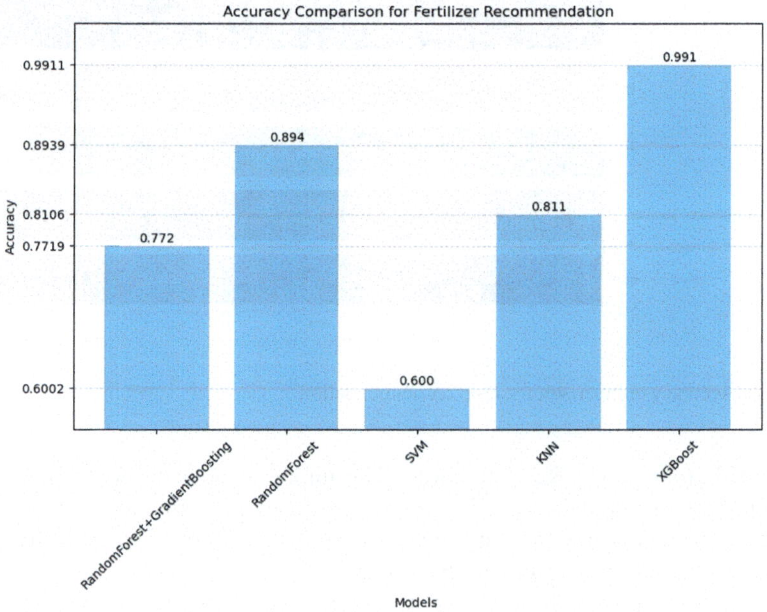

Fig. 10.5 Fertilizer recommendation

(K), and nitrogen (N). Positive Correlations: A correlation between two variables is said to be positive if it is closer to 1. For instance, there appears to be a positive relationship between potassium (K) and nitrogen (N). This implies that potassium levels tend to rise in tandem with increases in nitrogen levels and vice versa. Negative Correlations: When two variables have a correlation coefficient that is closer to – 1, it is considered to be negatively correlated. The relationship between pH and potassium (K) seems to be somewhat inverse. This implies that potassium levels may marginally decline as pH rises (becomes more alkaline). Strength of Correlations: The heatmap's color intensity corresponds to the correlation's strength. In this case, greater correlations—whether positive or negative—are represented by deeper hues. Figure 10.6 gives the correlation amtrix.

The machine learning model's promising crop classification results are displayed in the Fig. 10.7. A number of crops, including turmeric, cotton, and sugarcane, attained almost flawless results in precision, recall, and F1-score. This demonstrates the model's remarkable capacity to accurately and confidently detect these crops. But there are certain issues that need to be resolved. It seems that the model has trouble with grapes. Although all anticipated grapes were accurately categorized (precision of 1.0), some real grape plantings may have been overlooked (lower recall of 0.80).

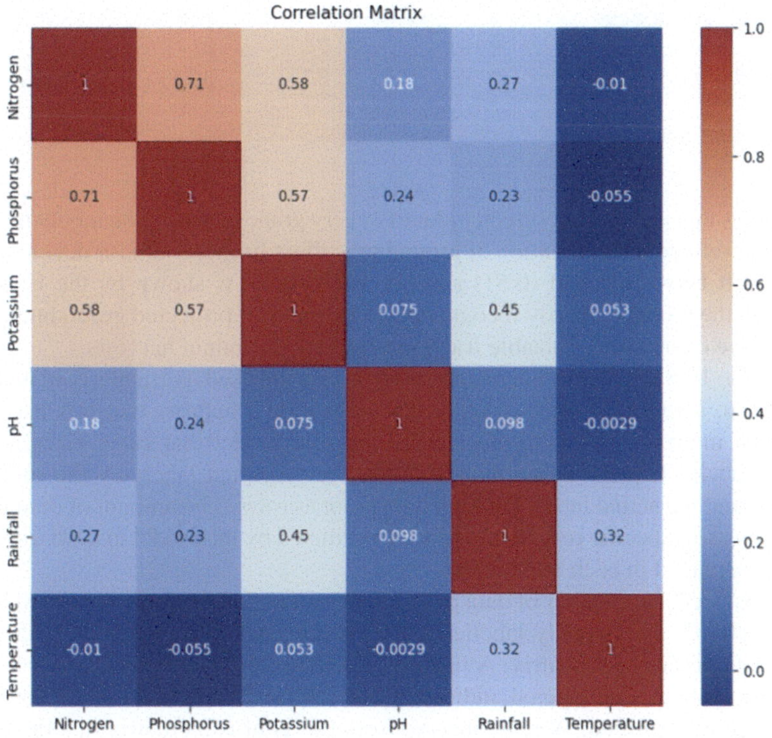

Fig. 10.6 Correlation matrix

Classification Report:

	precision	recall	f1-score	support
Cotton	1.00	0.96	0.98	140
Ginger	1.00	0.72	0.84	25
Gram	1.00	0.15	0.27	13
Grapes	1.00	0.80	0.89	5
Groundnut	0.86	0.51	0.64	49
Jowar	0.55	0.85	0.67	78
Maize	0.80	0.79	0.79	71
Masoor	1.00	1.00	1.00	4
Moong	0.55	0.35	0.43	17
Rice	1.00	1.00	1.00	64
Soybean	1.00	0.43	0.60	7
Sugarcane	0.95	1.00	0.97	123
Tur	1.00	0.86	0.93	22
Turmeric	1.00	1.00	1.00	3
Urad	1.00	0.40	0.57	15
Wheat	0.88	0.98	0.93	166
accuracy			0.87	802
macro avg	0.91	0.74	0.78	802
weighted avg	0.89	0.87	0.86	802

Fig. 10.7 Classification report

This shows that in order to correctly identify every grape instance, the model needs to be further refined. Classification of groundnuts offers further room for development. A trade-off between recall (0.51) and precision (0.86) is shown by the F1-score. Although the model correctly recognized the majority of projected groundnut crops, it may have overlooked a sizable fraction of actual groundnut harvests.

The Fig. 10.8 is a confusion matrix, which is a table used, particularly in machine learning, to assess how well an algorithm performs. It appears that this particular confusion matrix is assessing a machine learning model that categorizes sixteen distinct labels. The confusion matrix shows the projected labels by the model as columns and the actual labels for each data point as rows. The amount of data points that belong to a specific real label and were predicted by the model to have a specific label is displayed in each table cell.

For instance, the number of data points that were both accurately predicted by the model to be"0" and actually labelled as "0" is represented by the value 135 in the upper left corner of the matrix. A flawless model would ideally have all of its data points on the matrix's diagonal, indicating that each and every one of the data points was correctly identified. We can observe from this confusion matrix that the model does well for certain classes but not for others. By identifying label "0" correctly 135 times, the model appears to perform well.

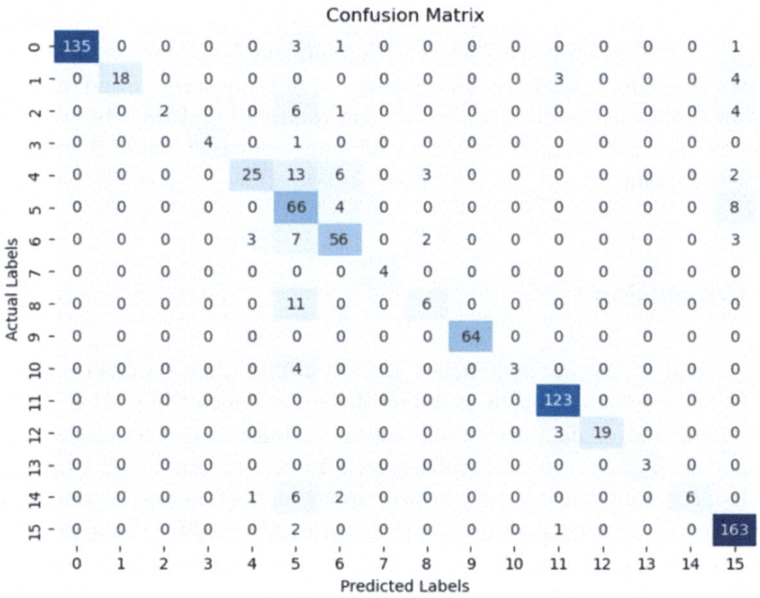

Fig. 10.8 Confusion matrix

10.6 Future Scope

A strong foundation for utilizing machine learning methods to assist sustainable farming practices is provided by the GrowWise platform. Although the current study shows that the generated models have the ability to provide precise guidance for fertilizer administration and crop selection, there are a number of directions that might be pursued for future research and development to further improve the platform's capabilities. Including Other Data Sources: The present research is based on a large dataset that includes data on climate, and soil properties. The accuracy and applicability of the recommendations could be greatly increased by incorporating data from other sources, such as real-time sensor data, meteorological forecasts, and remote sensing data. Meteorological forecasts could facilitate proactive decision-making, while remote sensing data could offer insightful information about crop growth patterns, soil moisture, and the health of foliage.

Integration of Market and Economic Factors: The main emphasis of the current study is on crop selection and fertilizer recommendations derived from agricultural and environmental factors. On the other hand, including market and economic variables—like crop prices, supply chain dynamics, and demand patterns—might give farmers a more thorough framework for making decisions. Through this integration, farmers would be able to make well-informed decisions from both an agricultural and economic perspective, which might lead to an increase in their resilience and profitability.

Continuous Model Refinement and Adaptation: It is essential to continuously refine and adapt the generated models as agricultural methods, environmental factors, and crop types change over time. The platform would stay accurate and relevant over time by including feedback mechanisms and routinely updating the models with fresh information and insights, which would represent the dynamic character of the agricultural domain.

10.7 Conclusion

Utilizing machine learning approaches, the GrowWise platform tackles the crucial issues of sustainable crop selection and fertilizer recommendation. This study highlights the potential of data-driven approaches in improving agricultural practices and enhancing food security by utilizing an extensive dataset and a strict methodology. First, Random Forest is the best-performing model for crop recommendations, according to our results, with an amazing accuracy of 0.869. This demonstrates Random Forest's resilience and dependability in identifying intricate associations in agricultural datasets and producing precise forecasts that are suited to farmers' requirements. Second, XGBoost performs better than other methods in forecasting nutrient requirements, with average forecasts that are closest to the actual amounts needed. This demonstrates how XGBoost may be used to combine gradient boosting and regularization strategies to maximize prediction accuracy, especially when crop growth nutrient optimization is involved.

Furthermore, XGBoost achieves the best accuracy of 0.991 according to our examination of fertilizer recommendations. According to this, XGBoost does a great job of understanding the complex interactions that exist between crop nutrient requirements, soil properties, and fertilizer compositions. This makes fertilizer application recommendations accurate and dependable.

The results of this study show that the GrowWise platform is feasible and efficient in giving farmers data-driven advice on which crops to plant and how much fertilizer to use. Farmers may enhance crop yields, preserve the environment, and ensure food security by implementing sustainable agricultural methods, optimizing resource utilization, and making well-informed decisions.

References

Adamchuk Viacheslav I et al (2004) On-the-go soil sensors for precision agriculture. Comput Electron Agricult 44(1):71–91

Akshatha KR, Shreedhara KS (2018) Implementation of machine learning algorithms for crop recommendation using precision agriculture. Int J Res Eng Sci Manag (IJRESM) 1(6):58–60

Bellis Emily S et al (2022) Detecting intra-field variation in rice yield with unmanned aerial vehicle imagery and deep learning. Front Plant Sci 13:716506

Chana AM, Batchakui B, Nges BB (2023) Real-time crop prediction based on soil fertility and weather forecast using IoT and a machine learning algorithm. Agricult Sci 14(5):645–664
Durai Arumugam SSL, Praveenkumar R, Balaji V (2023) An intelligent crop recommendation system using deep learning. Int J Intell Syst Appl Eng 11(10s):423–428
Ennaji O, Vergutz L, El Allali A (2023) Machine learning in nutrient management: a review. Art Intell Agricult 9:1–11
Kamilaris A, Prenafeta-Boldú FX (2018) Deep learning in agriculture: a survey. Comput Electron Agric 147:70–90
Macharia M et al (2016) Fertilizer use optimization approach: an innovation to increase agricultural profitability for African farmers. Afr J Agric Res 11(38):3587–3597
Mohammad R, Kodandaramu Ch, Achyuta Rao A (2022) A machine learning method for crop recommendation. J Eng Sci 13(10):51–64
Praynlin E, Nithin H, Tabassum N, Nikitha KN, Nikitha Y (2023) Machine learning approach for crop and fertilizer recommendation. Int J Adv Eng Archit Sci Technol 1:36–44
Priyadharshini A et al (2021) Intelligent crop recommendation system using machine learning. In: 5th international conference on computing methodologies and communication (ICCMC). IEEE
Rama Krishna S, Amrutha K (2020) A novel ensembling technique for crop recommendation system using machine learning. Int J Adv Sci Technol 29(5):874–882
Rajakumaran M et al (2024) Crop yield prediction using multi-attribute weighted tree-based support vector machine. Meas Sens 31:101002
Rohit Kumar R et al (2017) Crop recommendation system to maximize crop yield using machine learning technique. Int Res J Eng Technol 4(12):950–953
Sambandh Bhusan D et al (2022) Nutrient optimization for plant growth in Aquaponic irrigation using machine learning for small training datasets. Art Intell Agricult 6:68–76
Suchithra MS, Pai ML (2020) Improving the prediction accuracy of soil nutrient classification by optimizing extreme learning machine parameters. Inform Proc Agricult 7(1):72–82
Zhu N et al (2018) Deep learning for smart agriculture: concepts, tools, applications, and opportunities. Int J Agricult Biol Eng 11(4):32–44

The manufacturer's authorised representative in the EU is Springer Nature Customer Service Centre GmbH, Europaplatz 3, 69115 Heidelberg, Germany. If you have any concerns regarding our products, please contact ProductSafety@springernature.com

Printed and bound by CPI Group (UK) Ltd, Croydon, CR0 4YY
01/10/2026
02225944-0017